British Railw

CW00591659

ELE
MULTIPLE UNITS

TWENTY-EIGHTH EDITION
2015

The Complete Guide to all
Electric Multiple Units which operate on
the national railway network

Robert Pritchard

ISBN 978 1909 431 15 7

© 2014. Platform 5 Publishing Ltd, 52 Broadfield Road, Sheffield, S8 0XJ,
England.

Printed in England by Berforts Information Press, Eynsham, Oxford.

CONTENTS

Light Rail & Metro Systems. Because of the large number of new EMUs that have been included in this year's book we have moved the Light Rail & Metro Systems section into BR Pocket Book No. 3: Diesel Multiple Units.

PROVISION OF INFORMATION

This book has been compiled with care to be as accurate as possible, but in some cases information is not easily available and the publisher cannot be held responsible for any errors or omissions. We would like to thank the companies and individuals which have been co-operative in supplying information to us. The authors of this series of books are always pleased to receive notification from readers of any inaccuracies readers may find in the series, to enhance future editions. Please send comments to:

Robert Pritchard, Platform 5 Publishing Ltd, 52 Broadfield Road, Sheffield, S8 0XJ, England.

e-mail: robert@platform5.com   **Tel:** 0114 255 2625   **Fax:** 0114 255 2471.

This book is updated to information received by 6 October 2014.

UPDATES

This book is updated to the Stock Changes given in **Today's Railways UK 155** (November 2014). Readers are therefore advised to update this book from the official Platform 5 Stock Changes published every month in **Today's Railways UK** magazine, starting with issue 156.

The Platform 5 magazine Today's Railways UK contains news and rolling stock information on the railways of Great Britain and Ireland and is published on the second Monday of every month. For further details of Today's Railways UK, please see the advertisement on the back cover of this book.

Front cover photograph: TransPennine Express 350 408 heads south at Abington with the 14.10 Edinburgh–Manchester Airport on 18/06/14.   **Andrew Wills**

BRITAIN'S RAILWAY SYSTEM

INFRASTRUCTURE & OPERATION

Britain's national railway infrastructure is owned by a "not for dividend" company, Network Rail. In September 2014 Network Rail was classified as a public sector company, being described by the Government as a "public sector arm's-length body of the Department for Transport".

Many stations and maintenance depots are leased to and operated by Train Operating Companies (TOCs), but some larger stations remain under Network Rail control. The only exception is the infrastructure on the Isle of Wight: Island Line was the only franchise that included the maintenance of the infrastructure as well as the operation of passenger services. As Island Line is now part of the South West Trains franchise, both the infrastructure and trains are operated by South West Trains.

Trains are operated by TOCs over Network Rail, regulated by access agreements between the parties involved. In general, TOCs are responsible for the provision and maintenance of the locomotives, rolling stock and staff necessary for the direct operation of services, whilst Network Rail is responsible for the provision and maintenance of the infrastructure and also for staff to regulate the operation of services.

The Department for Transport is the franchising authority for the national network, with Transport Scotland overseeing the award of the ScotRail franchise. Railway Franchise holders (TOCs) can take commercial risks, although some franchises are known as "management contracts", where ticket revenues pass directly to the DfT. Concessions (such as London Overground) see the operator paid a fee to run the service, usually within tightly specified guidelines. Operators running a Concession would not normally take commercial risks, although there are usually penalties and rewards in the contract.

During 2012 the letting of new franchises was suspended pending a review of the franchise system. The process was restarted in 2013 but it is going to take a number of years to catch-up and several franchises are receiving short-term extensions (or "Direct Awards") in the meantime.

DOMESTIC PASSENGER TRAIN OPERATORS

The large majority of passenger trains are operated by the TOCs on fixed-term franchises. Franchise expiry dates are shown in the list of franchisees below:

Franchise	Franchisee	Trading Name
Chiltern	Deutsche Bahn (Arriva) (until 31 December 2021)	Chiltern Railways
Cross-Country	Deutsche Bahn (Arriva) (until 31 March 2016)*	CrossCountry

Franchise extension to be negotiated to November 2019.

East Midlands	Stagecoach Group plc (until 31 March 2015)*	**East Midlands Trains**

Franchise extension to be negotiated to October 2017.

Essex Thameside	National Express Group plc (until 8 November 2029)	**c2c**
Greater Western	First Group plc (until 20 September 2015)	**First Great Western**

Franchise extension currently being negotiated.

Greater Anglia	Abellio (NS) (until 19 October 2016)	**Abellio Greater Anglia**
Integrated Kent	Govia Ltd (Go-Ahead/Keolis) (until 24 June 2018)	**Southeastern**
InterCity East Coast	Directly Operated Railways (until 31 March 2015)	**East Coast**

Currently run on an interim basis by DfT management company Directly Operated Railways (trading as East Coast). This arrangement is due to continue until a new franchise is let to the private sector, with the new franchise currently planned to start on 1 April 2015.

InterCity West Coast	Virgin Rail Group Ltd (until 31 March 2017)	**Virgin Trains**
London Rail	MTR/Deutsche Bahn (until 12 November 2016)	**London Overground**

This is a Concession and is different from other rail franchises, as fares and service levels are set by Transport for London instead of the DfT.

Merseyrail Electrics	Serco/Abellio (NS) (until 19 July 2028)	**Merseyrail**

Under the control of Merseytravel PTE instead of the DfT. Franchise reviewed every five years to fit in with the Merseyside Local Transport Plan.

Northern Rail	Serco/Abellio (NS) (until 1 February 2016)	**Northern**
ScotRail	First Group plc (until 31 March 2015)	**ScotRail**

Abellio has won the contract to operate the ScotRail franchise from April 2015.

South Central	Govia Ltd (Go-Ahead/Keolis) (until 25 July 2015)	**Southern**

Upon termination of the Southern franchise in July 2015 it is to be combined with the new Thameslink, Southern & Great Northern franchise (also operated by Govia).

South Western	Stagecoach Group plc (until 3 February 2017)*	**South West Trains**

Franchise extension to be negotiated to April 2019.

Thameslink & Great Northern	First Group plc (until 19 September 2021)	**Govia Thameslink Railway**
Trans-Pennine Express	First Group/Keolis (until 1 April 2015)	**TransPennine Express**

Franchise extension to be negotiated to February 2016.

Wales & Borders	Deutsche Bahn (Arriva) (until 14 October 2018)*	**Arriva Trains Wales**

The franchise agreement includes the provision for the term to be further extended by mutual agreement by up to five years beyond October 2018. Management of the franchise is devolved to the Welsh Government, but DfT is still the procuring authority.

West Midlands	Govia Ltd (Go-Ahead/Keolis) (until 19 September 2015)*	**London Midland**

Franchise extension to be negotiated to June 2017.

* Franchise agreement includes provision for an extension of up to seven 4-week reporting periods.

The following operators run non-franchised services (* special summer services only):

Operator	Trading Name	Route
BAA	Heathrow Express	London Paddington–Heathrow Airport
First Hull Trains	First Hull Trains	London King's Cross–Hull
Grand Central	Grand Central	London King's Cross–Sunderland/ Bradford Interchange
North Yorkshire Moors Railway Enterprises	North Yorkshire Moors Railway	Pickering–Grosmont–Whitby/ Battersby
West Coast Railway Company	West Coast Railway Company	Birmingham–Stratford-upon-Avon* Fort William–Mallaig* York–Wakefield–York–Scarborough*

INTERNATIONAL PASSENGER OPERATIONS

Eurostar operates passenger services between the UK and mainland Europe.

Eurostar International, established in 2010, is jointly owned by the SNCF (the national operator of France, 55%), SNCB (the national operator of Belgium, 5%) and HM Treasury (40%). The 40% UK stake was transferred from London & Continental Railways (LCR) to HM Treasury in 2014. LCR had bought British Rail's interest in Eurostar at the time of the UK railway privatisation in 1996.

In addition, a service for the conveyance of accompanied road vehicles through the Channel Tunnel is provided by the tunnel operating company, Eurotunnel.

FREIGHT TRAIN OPERATIONS

The following operators operate freight services or empty passenger stock workings under "Open Access" arrangements:

Colas Rail
DB Schenker Rail (UK)
Devon & Cornwall Railways
Direct Rail Services (DRS)
Freightliner
GB Railfreight (owned by Eurotunnel)
West Coast Railway Company

INTRODUCTION

CLASSIFICATION

All dimensions and weights are quoted for vehicles in an "as new" condition with all necessary supplies on board. Dimensions are quoted in the order length x overall width. All lengths quoted are over buffers or couplers as appropriate. Where two lengths are quoted, the first refers to outer vehicles in a set and the second to inner vehicles. All weights are shown as metric tonnes (t = tonnes).

Bogie Types are quoted in the format motored/non-motored (eg BP20/BT13 denotes BP20 motored bogies and BT non-motored bogies).

Unless noted to the contrary, all vehicles listed have bar couplers at non-driving ends.

Traction motors power details refer to each motored car per unit.

Vehicles ordered under the auspices of BR were allocated a Lot (batch) number when ordered and these are quoted in class headings and sub-headings. Vehicles ordered since 1995 have no Lot Numbers, but the manufacturer and location that they were built is given.

LAYOUT OF INFORMATION

25 kV AC 50 Hz overhead Electric Multiple Units (EMUs) and dual voltage EMUs are listed in numerical order of set numbers. Individual "loose" vehicles are listed in numerical order after vehicles formed into fixed formations.

750 V DC third rail EMUs are listed in numerical order of class number, then in numerical order of set number. Some of these use the former Southern Region four-digit set numbers. These are derived from theoretical six digit set numbers which are the four-digit set number prefixed by the first two numbers of the class.

Where sets or vehicles have been renumbered in recent years, former numbering detail is shown alongside current detail. Each entry is laid out as in the following example:

Set No.	Detail	Livery	Owner	Operator	Allocation	Formation			
315803	†	**GA**	E	*GA*	IL	64465	71283	71391	64466

Detail Differences. Only detail differences which currently affect the areas and types of train which vehicles may work are shown. All other detail differences are excluded. Where such differences occur within a class or part class, these are shown alongside the individual set or vehicle number. Meaning of abbreviations are detailed in individual class headings.

Set Formations. Set formations shown are those normally maintained. Readers should note some set formations might be temporarily varied from time to time to suit maintenance and/or operational requirements.

Codes. Codes are used to denote the livery, owner, operator and depot of each unit. Details of these will be found in Section 8 of this book. Where a unit or spare car is off-lease, the operator column will be left blank.

Names. Only names carried with official sanction are listed. Names are shown in UPPER/lower case characters as actually shown on the name carried on the vehicle(s). Unless otherwise shown, complete units are regarded as named rather than just the individual car(s) which carry the name.

GENERAL INFORMATION

NUMBERING

25 kV AC 50 Hz overhead and "Versatile" EMUs are classified in the series 300–399.

750 V DC third rail EMUs are classified in the series 400–599.

Service units are classified in the series 900–949.

EMU individual cars are numbered in the series 61000–78999, except for vehicles used on the Isle of Wight – which are numbered in a separate series, and the Class 378s, 380s and 395s, which take up 38xxx and 39xxx series'.

Any vehicle constructed or converted to replace another vehicle following accident damage and carrying the same number as the original vehicle is denoted by the suffix[II] in this publication.

Principal details and dimensions are quoted for each class in metric and/or imperial units as considered appropriate bearing in mind common UK usage.

OPERATING CODES

These codes are used by train operating company staff to describe the various different types of vehicles and normally appear on data panels on the inner (ie non driving) ends of vehicles.

A "B" prefix indicates a battery vehicle.
A "P" prefix indicates a trailer vehicle on which is mounted the pantograph, instead of the default case where the pantograph is mounted on a motor vehicle.

The first part of the code describes whether or not the car has a motor or a driving cab as follows:

DM	Driving motor	M	Motor
DT	Driving trailer	T	Trailer

The next letter is a "B" for cars with a brake compartment.
This is followed by the saloon details:

F	First	S	Standard
C	Composite		

The next letter denotes the style of accommodation as follows:

O	Open	K	Side compartment with lavatory
so	Semi-open		(part compartments, part open). All other vehicles are assumed to consist solely of open saloons.

Finally vehicles with a buffet or kitchen area are suffixed RB or RMB for a miniature buffet counter.

Where two vehicles of the same type are formed within the same unit, the above codes may be suffixed by (A) and (B) to differentiate between the vehicles.

A composite is a vehicle containing both First and Standard Class accommodation, whilst a brake vehicle is a vehicle containing separate specific accommodation for the conductor.

Special Note: Where vehicles have been declassified, the correct operating code which describes the actual vehicle layout is quoted in this publication.

The following codes are used to denote special types of vehicle:

DMLF   Driving Motor Lounge First
DMLV   Driving Motor Luggage Van
MBRBS   Motor buffet standard with luggage space and guard's compartment.
TFH   Trailer First with Handbrake

BUILD DETAILS

Lot Numbers
Vehicles ordered under the auspices of BR were allocated a Lot (batch) number when ordered and these are quoted in class headings and sub-headings.

ACCOMMODATION

The information given in class headings and sub-headings is in the form F/S nT (or TD) nW. For example 12/54 1T 1W denotes 12 First Class and 54 Standard Class seats, one toilet and one space for a wheelchair. A number in brackets (i.e. (2)) denotes tip-up seats (in addition to the fixed seats). Tip-up seats in vestibules do not count. The seating layout of open saloons is shown as 2+1, 2+2 or 3+2 as the case may be. Where units have first class accommodation as well as standard and the layout is different for each class then these are shown separately prefixed by "1:" and "2:". Compartments are three seats a side in First Class and mostly four a side in Standard Class in EMUs. TD denotes a toilet suitable for use by a disabled person.

ABBREVIATIONS

The following standard abbreviations are used in class headings and also throughout this publication:

AC	Alternating Current.	kW	kilowatts.
BR	British Railways.	LT	London Transport.
BSI	Bergische Stahl Industrie.	LUL	London Underground Limited.
DC	Direct Current.	m	metres.
EMU	Electric Multiple Unit.	mph	miles per hour.
Hz	Hertz.	SR	BR Southern Region.
kN	kilonewtons.	t	tonnes.
km/h	kilometres per hour.	V	volts.

1. 25 kV AC 50 Hz OVERHEAD & DUAL VOLTAGE UNITS

Except where otherwise stated, all units in this section operate on 25 kV AC 50 Hz overhead only.

CLASS 313   BREL YORK

Inner suburban units.

Formation: DMSO–PTSO–BDMSO or DMSO–TSO–BDMSO.
Systems: 25 kV AC overhead/750 V DC third rail.
Construction: Steel underframe, aluminium alloy body and roof.
Traction Motors: Four GEC G310AZ of 82.125 kW.
Wheel Arrangement: Bo-Bo + 2-2 + Bo-Bo.
Braking: Disc & rheostatic.   **Dimensions:** 20.33/20.18 x 2.82 m.
Bogies: BX1.   **Couplers:** Tightlock.
Gangways: Within unit + end doors.   **Control System:** Camshaft.
Doors: Sliding.   **Maximum Speed:** 75 mph.
Seating Layout: Various, see sub-class headings.
Multiple Working: Within class.

DMSO. Lot No. 30879 1976–77. –/74. 36.0 t.
PTSO. Lot No. 30880 1976–77. –/83. 31.0 t.
BDMSO. Lot No. 30885 1976–77. –/74. 37.5 t.

Class 313/0. Standard Design. Refurbished with high back seats (3+2 facing).

313018	**FU**	E	*GT*	HE	62546	71230	62610
313024	**FU**	E	*GT*	HE	62552	71236	62616
313025	**FU**	E	*GT*	HE	62553	71237	62617
313026	**FU**	E	*GT*	HE	62554	71238	62618
313027	**FU**	E	*GT*	HE	62555	71239	62619
313028	**FU**	E	*GT*	HE	62556	71240	62620
313029	**FU**	E	*GT*	HE	62557	71241	62621
313030	**FU**	E	*GT*	HE	62558	71242	62622
313031	**FU**	E	*GT*	HE	62559	71243	62623
313032	**FU**	E	*GT*	HE	62560	71244	62643
313033	**FU**	E	*GT*	HE	62561	71245	62625
313035	**FU**	E	*GT*	HE	62563	71247	62627
313036	**FU**	E	*GT*	HE	62564	71248	62628
313037	**FU**	E	*GT*	HE	62565	71249	62629
313038	**FU**	E	*GT*	HE	62566	71250	62630
313039	**FU**	E	*GT*	HE	62567	71251	62631
313040	**FU**	E	*GT*	HE	62568	71252	62632
313041	**FU**	E	*GT*	HE	62569	71253	62633
313042	**FU**	E	*GT*	HE	62570	71254	62634
313043	**FU**	E	*GT*	HE	62571	71255	62635
313044	**FU**	E	*GT*	HE	62572	71256	62636
313045	**FU**	E	*GT*	HE	62573	71257	62637
313046	**FU**	E	*GT*	HE	62574	71258	62638

313047	**FU**	E	*GT*	HE	62575	71259	62639
313048	**FU**	E	*GT*	HE	62576	71260	62640
313049	**FU**	E	*GT*	HE	62577	71261	62641
313050	**FU**	E	*GT*	HE	62578	71262	62649
313051	**FU**	E	*GT*	HE	62579	71263	62624
313052	**FU**	E	*GT*	HE	62580	71264	62644
313053	**FU**	E	*GT*	HE	62581	71265	62645
313054	**FU**	E	*GT*	HE	62582	71266	62646
313055	**FU**	E	*GT*	HE	62583	71267	62647
313056	**FU**	E	*GT*	HE	62584	71268	62648
313057	**FU**	E	*GT*	HE	62585	71269	62642
313058	**FU**	E	*GT*	HE	62586	71270	62650
313059	**FU**	E	*GT*	HE	62587	71271	62651
313060	**FU**	E	*GT*	HE	62588	71272	62652
313061	**FU**	E	*GT*	HE	62589	71273	62653
313062	**FU**	E	*GT*	HE	62590	71274	62654
313063	**FU**	E	*GT*	HE	62591	71275	62655
313064	**FU**	E	*GT*	HE	62592	71276	62656

Name (carried on PTSO): 313054   Captain William Leefe Robinson V.C.

Class 313/1. Former London Overground units. Original low back seats (3+2 facing). Details as Class 313/0.

313122	**FU**	E	*GT*	HE	62550	71234	62614
313123	**FU**	E	*GT*	HE	62551	71235	62615
313134	**FU**	E	*GT*	HE	62562	71246	62626

Names (carried on PTSO):

313122   Eric Roberts 1946–2012 "The Flying Nottsman"
313134   City of London

Class 313/2. Southern units. Units refurbished for Southern for Brighton Coastway services. Fitted with 2+2 mainly facing high-back seating. 750V DC only (pantographs removed).

DMSO. Lot No. 30879 1976–77. –/64. 36.0 t.
TSO. Lot No. 30880 1976–77. –/68.   . t.
BDMSO. Lot No. 30885 1976–77. –/64. 37.5 t.

313201	(313101)	**SN**	PC	*SN*	BI	62529	71213	62593
313202	(313102)	**SN**	PC	*SN*	BI	62530	71214	62594
313203	(313103)	**SN**	PC	*SN*	BI	62531	71215	62595
313204	(313104)	**SN**	PC	*SN*	BI	62532	71216	62596
313205	(313105)	**SN**	PC	*SN*	BI	62533	71217	62597
313206	(313106)	**SN**	PC	*SN*	BI	62534	71218	62598
313207	(313107)	**SN**	PC	*SN*	BI	62535	71219	62599
313208	(313108)	**SN**	PC	*SN*	BI	62536	71220	62600
313209	(313109)	**SN**	PC	*SN*	BI	62537	71221	62601
313210	(313110)	**SN**	PC	*SN*	BI	62538	71222	62602
313211	(313111)	**SN**	PC	*SN*	BI	62539	71223	62603
313212	(313112)	**SN**	PC	*SN*	BI	62540	71224	62604
313213	(313113)	**SN**	PC	*SN*	BI	62541	71225	62605
313214	(313114)	**SN**	PC	*SN*	BI	62542	71226	62606

313215	(313115)	**SN**	PC	*SN*	BI	62543	71227	62607
313216	(313116)	**SN**	PC	*SN*	BI	62544	71228	62608
313217	(313117)	**SN**	PC	*SN*	BI	62545	71229	62609
313219	(313119)	**SN**	PC	*SN*	BI	62547	71231	62611
313220	(313120)	**SN**	PC	*SN*	BI	62548	71232	62612

CLASS 314   BREL YORK

Inner suburban units.

Formation: DMSO–PTSO–DMSO.
Construction: Steel underframe, aluminium alloy body and roof.
Traction Motors: Four GEC G310AZ (* Brush TM61-53) of 82.125 kW.
Wheel Arrangement: Bo-Bo + 2-2 + Bo-Bo.
Braking: Disc & rheostatic.   **Dimensions:** 20.33/20.18 x 2.82 m.
Bogies: BX1.   **Couplers:** Tightlock.
Gangways: Within unit + end doors. **Control System:** Thyristor.
Doors: Sliding.   **Maximum Speed:** 70 mph.
Seating Layout: 3+2 low-back facing.
Multiple Working: Within class and with Class 315.

DMSO. Lot No. 30912 1979. –/68. 34.5 t.
64588II. DMSO. Lot No. 30908 1978–80. Rebuilt Railcare Glasgow 1996 from
Class 507 No. 64426. The original 64588 was scrapped. –/74. 34.5 t.
PTSO. Lot No. 30913 1979. –/76. 33.0 t.

314201	*	**SC**	A	*SR*	GW	64583	71450	64584
314202	*	**SC**	A	*SR*	GW	64585	71451	64586
314203	*	**SR**	A	*SR*	GW	64587	71452	64588II
314204	*	**SR**	A	*SR*	GW	64589	71453	64590
314205	*	**SC**	A	*SR*	GW	64591	71454	64592
314206	*	**SC**	A	*SR*	GW	64593	71455	64594
314207		**SC**	A	*SR*	GW	64595	71456	64596
314208		**SR**	A	*SR*	GW	64597	71457	64598
314209		**SC**	A	*SR*	GW	64599	71458	64600
314210		**SC**	A	*SR*	GW	64601	71459	64602
314211		**SR**	A	*SR*	GW	64603	71460	64604
314212		**SR**	A	*SR*	GW	64605	71461	64606
314213		**SC**	A	*SR*	GW	64607	71462	64608
314214		**SC**	A	*SR*	GW	64609	71463	64610
314215		**SC**	A	*SR*	GW	64611	71464	64612
314216		**SC**	A	*SR*	GW	64613	71465	64614

CLASS 315   BREL YORK

Inner suburban units.

Formation: DMSO–TSO–PTSO–DMSO.
Construction: Steel underframe, aluminium alloy body and roof.
Traction Motors: Four Brush TM61-53 (* GEC G310AZ) of 82.125 kW.
Wheel Arrangement: Bo-Bo + 2-2 + 2-2 + Bo-Bo.
Braking: Disc & rheostatic.   **Dimensions:** 20.18 x 2.82 m.
Bogies: BX1.   **Couplers:** Tightlock.

Gangways: Within unit + end doors. **Control System:** Thyristor.
Doors: Sliding.   **Maximum Speed:** 75 mph.
Seating Layout: 3+2 low-back facing.
Multiple Working: Within class and with Class 314.

DMSO. Lot No. 30902 1980–81. –/74. 35.0 t († 38.2 t).
TSO. Lot No. 30904 1980–81. –/86. 25.5 t († 27.4 t).
PTSO. Lot No. 30903 1980–81. –/84 († –/75(7) 2W). 32.0 t († 33.8 t).

315801	†	**GA**	E	*GA*	IL	64461	71281	71389	64462
315802	†	**GA**	E	*GA*	IL	64463	71282	71390	64464
315803	†	**GA**	E	*GA*	IL	64465	71283	71391	64466
315804	†	**GA**	E	*GA*	IL	64467	71284	71392	64468
315805	†	**GA**	E	*GA*	IL	64469	71285	71393	64470
315806	†	**GA**	E	*GA*	IL	64471	71286	71394	64472
315807	†	**GA**	E	*GA*	IL	64473	71287	71395	64474
315808	†	**GA**	E	*GA*	IL	64475	71288	71396	64476
315809		**GA**	E	*GA*	IL	64477	71289	71397	64478
315810	†	**GA**	E	*GA*	IL	64479	71290	71398	64480
315811	†	**GA**	E	*GA*	IL	64481	71291	71399	64482
315812	†	**GA**	E	*GA*	IL	64483	71292	71400	64484
315813	†	**GA**	E	*GA*	IL	64485	71293	71401	64486
315814	†	**GA**	E	*GA*	IL	64487	71294	71402	64488
315815	†	**GA**	E	*GA*	IL	64489	71295	71403	64490
315816	†	**GA**	E	*GA*	IL	64491	71296	71404	64492
315817	†	**GA**	E	*GA*	IL	64493	71297	71405	64494
315818	†	**GA**	E	*GA*	IL	64495	71298	71406	64496
315819	†	**GA**	E	*GA*	IL	64497	71299	71407	64498
315820	†	**GA**	E	*GA*	IL	64499	71300	71408	64500
315821	†	**GA**	E	*GA*	IL	64501	71301	71409	64502
315822	†	**GA**	E	*GA*	IL	64503	71302	71410	64504
315823	†	**GA**	E	*GA*	IL	64505	71303	71411	64506
315824		**1**	E	*GA*	IL	64507	71304	71412	64508
315825	†	**GA**	E	*GA*	IL	64509	71305	71413	64510
315826	†	**GA**	E	*GA*	IL	64511	71306	71414	64512
315827	†	**GA**	E	*GA*	IL	64513	71307	71415	64514
315828	†	**GA**	E	*GA*	IL	64515	71308	71416	64516
315829	†	**GA**	E	*GA*	IL	64517	71309	71417	64518
315830	†	**GA**	E	*GA*	IL	64519	71310	71418	64520
315831	†	**GA**	E	*GA*	IL	64521	71311	71419	64522
315832	†	**GA**	E	*GA*	IL	64523	71312	71420	64524
315833		**1**	E	*GA*	IL	64525	71313	71421	64526
315834		**1**	E	*GA*	IL	64527	71314	71422	64528
315835		**1**	E	*GA*	IL	64529	71315	71423	64530
315836		**1**	E	*GA*	IL	64531	71316	71424	64532
315837		**1**	E	*GA*	IL	64533	71317	71425	64534
315838		**1**	E	*GA*	IL	64535	71318	71426	64536
315839		**1**	E	*GA*	IL	64537	71319	71427	64538
315840		**1**	E	*GA*	IL	64539	71320	71428	64540
315841	†	**GA**	E	*GA*	IL	64541	71321	71429	64542
315842	*	**1**	E	*GA*	IL	64543	71322	71430	64544
315843	*	**1**	E	*GA*	IL	64545	71323	71431	64546

315844	*†	**GA**	E	*GA*	IL	64547	71324	71432	64548
315845	*†	**GA**	E	*GA*	IL	64549	71325	71433	64550
315846	*†	**GA**	E	*GA*	IL	64551	71326	71434	64552
315847	*	**1**	E	*GA*	IL	64553	71327	71435	64554
315848	*	**1**	E	*GA*	IL	64555	71328	71436	64556
315849	*	**1**	E	*GA*	IL	64557	71329	71437	64558
315850	*	**1**	E	*GA*	IL	64559	71330	71438	64560
315851	*	**1**	E	*GA*	IL	64561	71331	71439	64562
315852	*	**1**	E	*GA*	IL	64563	71332	71440	64564
315853	*	**1**	E	*GA*	IL	64565	71333	71441	64566
315854	*†	**GA**	E	*GA*	IL	64567	71334	71442	64568
315855	*†	**GA**	E	*GA*	IL	64569	71335	71443	64570
315856	*	**1**	E	*GA*	IL	64571	71336	71444	64572
315857	*	**1**	E	*GA*	IL	64573	71337	71445	64574
315858	*†	**GA**	E	*GA*	IL	64575	71338	71446	64576
315859	*	**1**	E	*GA*	IL	64577	71339	71447	64578
315860	*	**1**	E	*GA*	IL	64579	71340	71448	64580
315861	*	**1**	E	*GA*	IL	64581	71341	71449	64582

Names (carried on DMSO):

315817	Transport for London
315829	London Borough of Havering Celebrating 40 years
315845	Herbie Woodward
315857	Stratford Connections

CLASS 317   BREL YORK/DERBY

Outer suburban units.

Formation: Various, see sub-class headings.
Construction: Steel.
Traction Motors: Four GEC G315BZ of 247.5 kW (except 317 722, see below).
Wheel Arrangement: 2-2 + Bo-Bo + 2-2 + 2-2.
Braking: Disc.   **Dimensions:** 19.83/20.18 x 2.82 m.
Bogies: BP20 (MSO), BT13 (others).   **Couplers:** Tightlock.
Gangways: Throughout   **Control System:** Thyristor.
Doors: Sliding.   **Maximum Speed:** 100 mph.
Seating Layout: Various, see sub-class headings.
Multiple Working: Within class & with Classes 318, 319, 320, 321, 322 and 323.

Class 317/1. Pressure ventilated.

Formation: DTSO–MSO–TCO–DTSO.
Seating Layout: 1: 2+2 facing, 2: 3+2 facing.

DTSO(A) Lot No. 30955 York 1981–82. –/74. 29.5 t.
MSO. Lot No. 30958 York 1981–82. –/79. 49.0 t.
TCO. Lot No. 30957 Derby 1981–82. 22/46 2T. 29.0 t.
DTSO(B) Lot No. 30956 York 1981–82. –/71. 29.5 t.

317337	**FU**	A	*GT*	HE	77036	62671	71613	77084
317338	**FU**	A	*GT*	HE	77037	62698	71614	77085
317339	**FU**	A	*GT*	HE	77038	62699	71615	77086

317340	**FU**	A	*GT*	HE	77039	62700	71616	77087
317341	**FU**	A	*GT*	HE	77040	62701	71617	77088
317342	**FU**	A	*GT*	HE	77041	62702	71618	77089
317343	**FU**	A	*GT*	HE	77042	62703	71619	77090
317344	**FU**	A	*GT*	HE	77029	62690	71620	77091
317345	**FU**	A	*GT*	HE	77044	62705	71621	77092
317346	**FU**	A	*GT*	HE	77045	62706	71622	77093
317347	**FU**	A	*GT*	HE	77046	62707	71623	77094
317348	**FU**	A	*GT*	HE	77047	62708	71624	77095

Names (carried on TCO):

317345 Driver John Webb | 317348 Richard A Jenner

Class 317/5. Pressure ventilated. Units renumbered from Class 317/1 in 2005 for West Anglia Metro services. Refurbished with new upholstery and Passenger Information Systems. Details as Class 317/1.

The original DTSO 77048 was written off after the Cricklewood accident of 1983. A replacement vehicle was built (at Wolverton) in 1987 and given the same number.

317501	**NX**	A	*GA*	IL	77024	62661	71577	77048
317502	**NX**	A	*GA*	IL	77001	62662	71578	77049
317503	**NX**	A	*GA*	IL	77002	62663	71579	77050
317504	**NX**	A	*GA*	IL	77003	62664	71580	77051
317505	**NX**	A	*GA*	IL	77004	62665	71581	77052
317506	**NX**	A	*GA*	IL	77005	62666	71582	77053
317507	**NX**	A	*GA*	IL	77006	62667	71583	77054
317508	**NX**	A	*GA*	IL	77010	62697	71587	77058
317509	**NX**	A	*GA*	IL	77011	62672	71588	77059
317510	**NX**	A	*GA*	IL	77012	62673	71589	77060
317511	**NC**	A	*GA*	IL	77014	62675	71591	77062
317512	**NC**	A	*GA*	IL	77015	62676	71592	77063
317513	**NX**	A	*GA*	IL	77016	62677	71593	77064
317514	**NX**	A	*GA*	IL	77017	62678	71594	77065
317515	**NX**	A	*GA*	IL	77019	62680	71596	77067

Name (carried on TCO):

317507 University of Cambridge 800 Years 1209–2009

Class 317/6. Convection heating. Units converted from Class 317/2 by Railcare, Wolverton 1998–99 with new Chapman seating.

Formation: DTSO–MSO–TSO–DTCO.
Seating Layout: 2+2 facing.

77200–219. DTSO. Lot No. 30994 York 1985–86. –/64. 29.5 t.
77280–283. DTSO. Lot No. 31007 York 1987. –/64. 29.5 t.
62846–865. MSO. Lot No. 30996 York 1985–86. –/70. 49.0 t.
62886–889. MSO. Lot No. 31009 York 1987. –/70. 49.0 t.
71734–753. TSO. Lot No. 30997 York 1985–86. –/62 2T. 29.0 t.
71762–765. TSO. Lot No. 31010 York 1987. –/62 2T. 29.0 t.
77220–239. DTCO. Lot No. 30995 York 1985–86. 24/48. 29.5 t.
77284–287. DTCO. Lot No. 31008 York 1987. 24/48. 29.5 t.

317649	**NC**	A	*GA*	IL	77200	62846	71734	77220
317650	**NC**	A	*GA*	IL	77201	62847	71735	77221
317651	**NC**	A	*GA*	IL	77202	62848	71736	77222
317652	**NC**	A	*GA*	IL	77203	62849	71739	77223
317653	**NC**	A	*GA*	IL	77204	62850	71738	77224
317654	**NC**	A	*GA*	IL	77205	62851	71737	77225
317655	**GA**	A	*GA*	IL	77206	62852	71740	77226
317656	**NC**	A	*GA*	IL	77207	62853	71742	77227
317657	**NC**	A	*GA*	IL	77208	62854	71741	77228
317658	**GA**	A	*GA*	IL	77209	62855	71743	77229
317659	**GA**	A	*GA*	IL	77210	62856	71744	77230
317660	**GA**	A	*GA*	IL	77211	62857	71745	77231
317661	**GA**	A	*GA*	IL	77212	62858	71746	77232
317662	**GA**	A	*GA*	IL	77213	62859	71747	77233
317663	**GA**	A	*GA*	IL	77214	62860	71748	77234
317664	**GA**	A	*GA*	IL	77215	62861	71749	77235
317665	**GA**	A	*GA*	IL	77216	62862	71750	77236
317666	**NC**	A	*GA*	IL	77217	62863	71752	77237
317667	**GA**	A	*GA*	IL	77218	62864	71751	77238
317668	**GA**	A	*GA*	IL	77219	62865	71753	77239
317669	**NC**	A	*GA*	IL	77280	62886	71762	77284
317670	**GA**	A	*GA*	IL	77281	62887	71763	77285
317671	**NC**	A	*GA*	IL	77282	62888	71764	77286
317672	**GA**	A	*GA*	IL	77283	62889	71765	77287

Name (carried on DTCO): 317654 Richard Wells

Class 317/7. Units converted from Class 317/1 by Railcare, Wolverton 2000 for Stansted Express services between London Liverpool Street and Stansted. Air conditioning. Fitted with luggage stacks. Displaced from Stansted services in 2011 by Class 379s.

* 317722 has received new Bombardier MJA 280-8 AC traction motors as part of an Angel trial. Two vehicles (77021 and 62682, now in **GA** livery) have also received an interior refurbishment with new Fainsa seating whilst the other two vehicles have been left in their former Stansted Express condition (and are still in **NX** livery). The unit returned to use with Greater Anglia as a demonstrator in 2014.

Formation: DTSO–MSO–TSO–DTCO.
Seating Layout: 1: 2+1 facing, 2: 2+2 facing.

DTSO Lot No. 30955 York 1981–82. –/52 + catering point. 31.4 t.
MSO. Lot No. 30958 York 1981–82. –/62 (* –/64). 51.3 t.
TSO. Lot No. 30957 Derby 1981–82. –/42(5) 1W 1T 1TD. 30.2 t.
DTCO Lot No. 30956 York 1981–82. 22/16 + catering point. 31.6 t.

317708	**GA**	A		ZG	77007	62668	71584	77055
317709	**NX**	A	*GA*	IL	77008	62669	71585	77056
317710	**NX**	A		ZI	77009	62670	71586	77057
317711	**GA**	A		ZI	77013	62674	71590	77061
317719	**NX**	A	*GA*	IL	77018	62679	71595	77066
317722	* **GA/NX**	A	*GA*	IL	77021	62682	71598	77069
317723	**NX**	A		ZI	77022	62683	71599	77070

▲ Southern uses a fleet of refurbished Class 313s on the Coastway routes from Brighton. On 06/08/14 313 208 calls at Hove with the 12.09 West Worthing–Brighton. **Peter Weber**

▼ Most Class 315s are now in Greater Anglia livery. 315 858/838 arrive at Stratford with the 17.24 Shenfield–London Liverpool Street on 29/04/14. **Robert Pritchard**

▲ Angel demonstrator 317 722 passes Manor Park with the 11.45 London Liverpool Street–Ilford ecs on 10/06/14. This unit carries Greater Anglia livery on the two leading vehicles and National Express livery on the other two cars. **Antony Guppy**

▼ In the new Govia Thameslink livery, 319 010 calls at Mitcham Junction with the 09.44 St Albans–Sutton on 12/07/14. **Chris Wilson**

▲ ScotRail Saltire-liveried 320 311 passes Craigenhill with the 14.23 Lanark–Dalmuir on 18/02/13. **Robin Ralston**

▼ Greater Anglia-liveried 321 441 leads the 17.10 Southend Victoria–London Liverpool Street into Stratford on 29/04/14. **Robert Pritchard**

▲ London Midland-liveried 323 214 and 323 217 (nearest camera) leave Five Ways with the 10.57 Redditch–Four Oaks on 15/02/14. **Robert Pritchard**

▼ Royal Mail EMU 325 015 passes Wandel with the 17.49 Shieldmuir–Warrington RMT on 24/04/14. **Robin Ralston**

▲ Heathrow Express units 332 011 and 332 013 pass Royal Oak with the 08.57 Heathrow Terminal 5–London Paddington on 21/05/14. **Antony Guppy**

▼ All Class 334s are now in the ScotRail Saltire livery. 334 020 leaves Coatbridge Sunnyside with the 11.25 Helensburgh Central–Edinburgh on 02/06/13.
Robert Pritchard

▲ London Midland 350 112 calls at Atherstone with the 13.02 Crewe–London Euston on 11/08/14. **Dave Gommersall**

▼ c2c-liveried 357 007 passes Shadwell, alongside the Docklands Light Railway, with the 12.30 London Fenchurch Street–Southend Central on 15/05/14.
David Palmer

▲ Heathrow Connect-liveried 360 201 leaves Southall with the 14.03 Hayes & Harlington–London Paddington on 23/08/14. **Antony Guppy**

▼ In the new Govia Thameslink livery, 365 517 brings up the rear of the 16.46 Peterborough–London King's Cross (led by First Capital Connect-liveried 365 501) at Hitchin on 09/05/14. **Mark Beal**

▲ 376 009 arrives at Beckenham Junction with the 11.28 special shuttle from Hayes on 06/04/12. **William Turvill**

▼ One of the new Southern 5-car 377s, dual voltage 377 702, passes Hemel Hempstead with the 12.13 Milton Keynes–South Croydon on 20/08/14. **Alisdair Anderson**

| 317729 | **GA** | A | | ZG | 77028 | 62689 | 71605 | 77076 |
| 317732 | **NX** | A | | ZI | 77031 | 62692 | 71608 | 77079 |

Names (carried on DTCO):

317709 Len Camp | 317723 The Tottenham Flyer

Class 317/8. Pressure Ventilated. Units refurbished and renumbered from Class 317/1 in 2005–06 at Wabtec, Doncaster for use on Stansted Express services. Displaced from Stansted services in 2011.

Formation: DTSO–MSO–TCO–DTSO.
Seating Layout: 1: 2+2 facing, 2: 3+2 facing.

DTSO(A) Lot No. 30955 York 1981–82. –/66. 29.5 t.
MSO. Lot No. 30958 York 1981–82. –/71. 49.0 t.
TCO. Lot No. 30957 Derby 1981–82. 20/42 2T. 29.0 t.
DTSO(B) Lot No. 30956 York 1981–82. –/66. 29.5 t.

317881	**NX**	A	*GA*	IL	77020	62681	71597	77068	
317882	**NC**	A	*GA*	IL	77023	62684	71600	77071	
317883	**NC**	A	*GA*	IL	77000	62685	71601	77072	
317884	**NC**	A	*GA*	IL	77025	62686	71602	77073	
317885	**NC**	A	*GA*	IL	77026	62687	71603	77074	
317886	**NC**	A	*GA*	IL	77027	62688	71604	77075	
317887	**NX**	A	*GA*	IL	77043	62704	71606	77077	
317888	**NX**	A	*GA*	IL	77030	62691	71607	77078	
317889	**NX**	A	*GA*	IL	77032	62693	71609	77080	
317890	**NX**	A	*GA*	IL	77033	62694	71610	77081	
317891	**NX**	A	*GA*	IL	77034	62695	71611	77082	
317892	**NX**	A	*GA*	IL	77035	62696	71612	77083	Ilford Depot

CLASS 318   BREL YORK

Outer suburban units. An overhaul programme is underway that is seeing a new universal access toilet (to comply with the 2020 accessibility regulations) fitted (units in **SR** livery).

Formation: DTSO–MSO–DTSO.
Construction: Steel.
Traction Motors: Four Brush TM 2141 of 268 kW.
Wheel Arrangement: 2-2 + Bo-Bo + 2-2.
Braking: Disc.
Bogies: BP20 (MSO), BT13 (others).
Gangways: Within unit.
Doors: Sliding.
Seating Layout: 3+2 facing.
Dimensions: 19.83/19.92 x 2.82 m.
Couplers: Tightlock.
Control System: Thyristor.
Maximum Speed: 90 mph.
Multiple Working: Within class & with Classes 317, 319, 320, 321, 322 and 323.

77240–259. DTSO. Lot No. 30999 1985–86. –/64 1T (–/53 1TD 2W). 30.0 t (* 32.9 t).
77288. DTSO. Lot No. 31020 1987. –/64 1T. 30.0 t.
62866–885. MSO. Lot No. 30998 1985–86. –/77 (*–/79). 50.9 t (* 53.0 t).
62890. MSO. Lot No. 31019 1987. –/77. 50.9 t.
77260–279. DTSO. Lot No. 31000 1985–86. –/72 (* –/74). 29.6 t (* 31.2 t).
77289. DTSO. Lot No. 31021 1987. –/72. 29.6 t.

318 250		**SC**	E	*SR*	GW	77240	62866	77260	
318 251	*	**SR**	E	*SR*	GW	77241	62867	77261	
318 252		**SC**	E	*SR*	GW	77242	62868	77262	
318 253		**SC**	E	*SR*	GW	77243	62869	77263	
318 254		**SC**	E	*SR*	GW	77244	62870	77264	
318 255		**SC**	E	*SR*	GW	77245	62871	77265	
318 256		**SC**	E	*SR*	GW	77246	62872	77266	
318 257	*	**SR**	E	*SR*	GW	77247	62873	77267	
318 258		**SC**	E	*SR*	GW	77248	62874	77268	
318 259	*	**SR**	E	*SR*	GW	77249	62875	77269	
318 260		**SC**	E	*SR*	GW	77250	62876	77270	
318 261		**SC**	E	*SR*	GW	77251	62877	77271	
318 262		**SC**	E	*SR*	GW	77252	62878	77272	
318 263		**SC**	E	*SR*	GW	77253	62879	77273	
318 264	*	**SR**	E	*SR*	GW	77254	62880	77274	
318 265		**SC**	E	*SR*	GW	77255	62881	77275	
318 266		**SC**	E	*SR*	GW	77256	62882	77276	STRATHCLYDER
318 267		**SC**	E	*SR*	GW	77257	62883	77277	
318 268		**SC**	E	*SR*	GW	77258	62884	77278	
318 269		**SC**	E	*SR*	GW	77259	62885	77279	
318 270		**SC**	E	*SR*	GW	77288	62890	77289	

CLASS 319   BREL YORK

Express and outer suburban units.

Formation: Various, see sub-class headings.
Systems: 25 kV AC overhead/750 V DC third rail.
Construction: Steel.
Traction Motors: Four GEC G315BZ of 268 kW.
Wheel Arrangement: 2-2 + Bo-Bo + 2-2 + 2-2.
Braking: Disc. **Dimensions:** 20.17/20.16 x 2.82 m.
Bogies: P7-4 (MSO), T3-7 (others). **Couplers:** Tightlock.
Gangways: Within unit + end doors. **Control System:** GTO chopper.
Doors: Sliding. **Maximum Speed:** 100 mph.
Seating Layout: Various, see sub-class headings.
Multiple Working: Within class & with Classes 317, 318, 320, 321, 322 and 323.

Class **319/0.** DTSO–MSO–TSO–DTSO.

Seating Layout: 3+2 facing.

DTSO(A). Lot No. 31022 (odd nos.) 1987–88. –/82. 28.2 t.
MSO. Lot No. 31023 1987–88. –/82. 49.2 t.
TSO. Lot No. 31024 1987–88. –/77 2T. 31.0 t.
DTSO(B). Lot No. 31025 (even nos.) 1987–88. –/78. 28.1 t.

319001	**TL**	P	*GT*	BF	77291	62891	71772	77290
319002	**FU**	P	*GT*	BF	77293	62892	71773	77292
319003	**FU**	P	*GT*	BF	77295	62893	71774	77294
319004	**TL**	P	*GT*	BF	77297	62894	71775	77296
319005	**TL**	P	*GT*	BF	77299	62895	71776	77298
319006	**FU**	P	*GT*	BF	77301	62896	71777	77300
319007	**FU**	P	*GT*	BF	77303	62897	71778	77302

319008	**SN**	P	*GT*	BF	77305	62898	71779	77304
319009	**TL**	P	*GT*	BF	77307	62899	71780	77306
319010	**TL**	P	*GT*	BF	77309	62900	71781	77308
319011	**TL**	P	*GT*	BF	77311	62901	71782	77310
319012	**SN**	P	*GT*	BF	77313	62902	71783	77312
319013	**SN**	P	*GT*	BF	77315	62903	71784	77314

Names (carried on TSO):

319001 Driver Mick Winnett	319011 John Ruskin College
319008 Cheriton	319013 The Surrey Hills
319009 Coquelles	

Class 319/2. DTSO–MSO–TSO–DTCO. Units converted from Class 319/0.

Seating Layout: 1: 2+1 facing, 2: 2+2 facing.

Advertising liveries: 319215 Visit Switzerland (red).
319 218 Lycamobile (white).

DTSO. Lot No. 31022 (odd nos.) 1987–88. –/64. 28.2 t.
MSO. Lot No. 31023 1987–88. –/73. 49.2 t.
TSO. Lot No. 31024 1987–88. –/52 1T 1TD. 31.0 t.
DTCO. Lot No. 31025 (even nos.) 1987–88. 18/36. 28.1 t.

319214	**SN**	P	*GT*	BF	77317	62904	71785	77316	
319215	**AL**	P	*GT*	BF	77319	62905	71786	77318	
319216	**SN**	P	*GT*	BF	77321	62906	71787	77320	
319217	**SN**	P	*GT*	BF	77323	62907	71788	77322	Brighton
319218	**AL**	P	*GT*	BF	77325	62908	71789	77324	Croydon
319219	**SN**	P	*GT*	BF	77327	62909	71790	77326	
319220	**SN**	P	*GT*	BF	77329	62910	71791	77328	

Class 319/3. DTSO–MSO–TSO–DTSO. Converted from Class 319/1 by replacing First Class seats with Standard Class seats. Used mainly on the Luton–Sutton/Wimbledon routes.

14 units (319 361–367/369/371/374–378) will transfer to Northern for use on newly electrified lines in the North-West in 2014–15.

Seating Layout: 3+2 facing.
Dimensions: 19.33 x 2.82 m.

DTSO(A). Lot No. 31063 1990. –/70. 29.0 t.
MSO. Lot No. 31064 1990. –/78. 50.6 t.
TSO. Lot No. 31065 1990. –/74 2T. 31.0 t.
DTSO(B). Lot No. 31066 1990. –/75. 29.7 t.

319361	**FU**	P	*NO*	AN	77459	63043	71929	77458
319362	**FU**	P	*NO*	AN	77461	63044	71930	77460
319363	**FU**	P	*NO*	AN	77463	63045	71931	77462
319364	**FU**	P	*GT*	BF	77465	63046	71932	77464
319365	**FU**	P	*GT*	BF	77467	63047	71933	77466
319366	**FU**	P	*GT*	BF	77469	63048	71934	77468
319367	**FU**	P	*GT*	BF	77471	63049	71935	77470
319368	**FU**	P	*GT*	BF	77473	63050	71936	77472
319369	**FU**	P	*GT*	BF	77475	63051	71937	77474

319370	**FU**	P	*GT*	BF	77477	63052	71938	77476
319371	**FU**	P	*GT*	BF	77479	63053	71939	77478
319372	**FU**	P	*GT*	BF	77481	63054	71940	77480
319373	**FU**	P	*GT*	BF	77483	63055	71941	77482
319374	**FU**	P	*GT*	BF	77485	63056	71942	77484
319375	**FU**	P	*GT*	BF	77487	63057	71943	77486
319376	**FU**	P	*GT*	BF	77489	63058	71944	77488
319377	**FU**	P	*GT*	BF	77491	63059	71945	77490
319378	**FU**	P	*GT*	BF	77493	63060	71946	77492
319379	**FU**	P	*GT*	BF	77495	63061	71947	77494
319380	**FU**	P	*GT*	BF	77497	63062	71948	77496
319381	**FU**	P	*GT*	BF	77973	63093	71979	77974
319382	**FU**	P	*GT*	BF	77975	63094	71980	77976
319383	**FU**	P	*GT*	BF	77977	63095	71981	77978
319384	**FU**	P	*GT*	BF	77979	63096	71982	77980
319385	**FU**	P	*GT*	BF	77981	63097	71983	77982
319386	**FU**	P	*GT*	BF	77983	63098	71984	77984

Name (carried on TSO):

319374 Bedford Cauldwell TMD

Class 319/4. DTCO–MSO–TSO–DTSO. Converted from Class 319/0. Refurbished with carpets. DTSO(A) converted to composite. Used mainly on the Bedford–Gatwick–Brighton route.

Seating Layout: 1: 2+1 facing 2: 2+2/3+2 facing.

77331–381. DTCO. Lot No. 31022 (odd nos.) 1987–88. 12/51. 28.2t.
77431–457. DTCO. Lot No. 31038 (odd nos.) 1988. 12/51. 28.2t.
62911–936. MSO. Lot No. 31023 1987–88. –/74. 49.2t.
62961–974. MSO. Lot No. 31039 1988. –/74. 49.2t.
71792–817. TSO. Lot No. 31024 1987–88. –/67 2T. 31.0t.
71866–879. TSO. Lot No. 31040 1988. –67 2T. 31.0t.
77330–380. DTSO. Lot No. 31025 (even nos.) 1987–88. –/71 1W. 28.1t.
77430–456. DTSO. Lot No. 31041 (even nos.) 1988. –/71 1W. 28.1t.

319421	**FU**	P	*GT*	BF	77331	62911	71792	77330
319422	**FU**	P	*GT*	BF	77333	62912	71793	77332
319423	**FU**	P	*GT*	BF	77335	62913	71794	77334
319424	**FU**	P	*GT*	BF	77337	62914	71795	77336
319425	**FU**	P	*GT*	BF	77339	62915	71796	77338
319426	**TL**	P	*GT*	BF	77341	62916	71797	77340
319427	**FU**	P	*GT*	BF	77343	62917	71798	77342
319428	**FU**	P	*GT*	BF	77345	62918	71799	77344
319429	**FU**	P	*GT*	BF	77347	62919	71800	77346
319430	**FU**	P	*GT*	BF	77349	62920	71801	77348
319431	**FU**	P	*GT*	BF	77351	62921	71802	77350
319432	**FU**	P	*GT*	BF	77353	62922	71803	77352
319433	**FU**	P	*GT*	BF	77355	62923	71804	77354
319434	**FU**	P	*GT*	BF	77357	62924	71805	77356
319435	**FU**	P	*GT*	BF	77359	62925	71806	77358
319436	**FU**	P	*GT*	BF	77361	62926	71807	77360
319437	**FU**	P	*GT*	BF	77363	62927	71808	77362

319438	**FU**	P	*GT*	BF	77365	62928	71809	77364
319439	**FU**	P	*GT*	BF	77367	62929	71810	77366
319440	**FU**	P	*GT*	BF	77369	62930	71811	77368
319441	**FU**	P	*GT*	BF	77371	62931	71812	77370
319442	**FU**	P	*GT*	BF	77373	62932	71813	77372
319443	**FU**	P	*GT*	BF	77375	62933	71814	77374
319444	**FU**	P	*GT*	BF	77377	62934	71815	77376
319445	**FU**	P	*GT*	BF	77379	62935	71816	77378
319446	**FU**	P	*GT*	BF	77381	62936	71817	77380
319447	**FU**	P	*GT*	BF	77431	62961	71866	77430
319448	**FU**	P	*GT*	BF	77433	62962	71867	77432
319449	**FU**	P	*GT*	BF	77435	62963	71868	77434
319450	**FU**	P	*GT*	BF	77437	62964	71869	77436
319451	**FU**	P	*GT*	BF	77439	62965	71870	77438
319452	**FU**	P	*GT*	BF	77441	62966	71871	77440
319453	**FU**	P	*GT*	BF	77443	62967	71872	77442
319454	**FU**	P	*GT*	BF	77445	62968	71873	77444
319455	**FU**	P	*GT*	BF	77447	62969	71874	77446
319456	**FU**	P	*GT*	BF	77449	62970	71875	77448
319457	**FU**	P	*GT*	BF	77451	62971	71876	77450
319458	**FU**	P	*GT*	BF	77453	62972	71877	77452
319459	**FU**	P	*GT*	BF	77455	62973	71878	77454
319460	**FU**	P	*GT*	BF	77457	62974	71879	77456

Names (carried on TSO):

319425	Transforming Travel
319435	Adrian Jackson-Robbins Chairman 1987–2007 Association of Public Transport Users
319444	City of St Albans
319446	St Pancras International
319448	Elstree Studios The Home of British Film and Television production
319449	King's Cross Thameslink

CLASS 320   BREL YORK

Suburban units. All refurbished 2011–13 and fitted with a new universal access toilet to comply with the 2020 accessibility regulations.

Formation: DTSO–MSO–DTSO.
Construction: Steel
Traction Motors: Four Brush TM2141B of 268 kW.
Wheel Arrangement: 2-2 + Bo-Bo + 2-2.
Braking: Disc.   **Dimensions:** 19.95 x 2.82 m.
Bogies: P7-4 (MSO), T3-7 (others).   **Couplers:** Tightlock.
Gangways: Within unit.   **Control System:** Thyristor.
Doors: Sliding.   **Maximum Speed:** 90 mph.
Seating Layout: 3+2 facing.
Multiple Working: Within class & with Classes 317, 318, 319, 321, 322 and 323.

DTSO (A). Lot No. 31060 1990. –/51(4) 1TD 2W. 31.7 t.
MSO. Lot No. 31062 1990. –/78. 52.6 t.
DTSO (B). Lot No. 31061 1990. –/77. 31.6 t.

320301	**SR**	E	*SR*	GW	77899	63021	77921
320302	**SR**	E	*SR*	GW	77900	63022	77922
320303	**SR**	E	*SR*	GW	77901	63023	77923
320304	**SR**	E	*SR*	GW	77902	63024	77924
320305	**SR**	E	*SR*	GW	77903	63025	77925
320306	**SR**	E	*SR*	GW	77904	63026	77926
320307	**SR**	E	*SR*	GW	77905	63027	77927
320308	**SR**	E	*SR*	GW	77906	63028	77928
320309	**SR**	E	*SR*	GW	77907	63029	77929
320310	**SR**	E	*SR*	GW	77908	63030	77930
320311	**SR**	E	*SR*	GW	77909	63031	77931
320312	**SR**	E	*SR*	GW	77910	63032	77932
320313	**SR**	E	*SR*	GW	77911	63033	77933
320314	**SR**	E	*SR*	GW	77912	63034	77934
320315	**SR**	E	*SR*	GW	77913	63035	77935
320316	**SR**	E	*SR*	GW	77914	63036	77936
320317	**SR**	E	*SR*	GW	77915	63037	77937
320318	**SR**	E	*SR*	GW	77916	63038	77938
320319	**SR**	E	*SR*	GW	77917	63039	77939
320320	**SR**	E	*SR*	GW	77918	63040	77940
320321	**SR**	E	*SR*	GW	77919	63041	77941
320322	**SR**	E	*SR*	GW	77920	63042	77942

CLASS 321   BREL YORK

Outer suburban units.

Formation: DTCO (DTSO on Class 321/9)–MSO–TSO–DTSO.
Construction: Steel.
Traction Motors: Four Brush TM2141C of 268 kW.
Wheel Arrangement: 2-2 + Bo-Bo + 2-2 + 2-2.
Braking: Disc.   **Dimensions:** 19.95 x 2.82 m.
Bogies: P7-4 (MSO), T3-7 (others).   **Couplers:** Tightlock.
Gangways: Within unit.   **Control System:** Thyristor.
Doors: Sliding.   **Maximum Speed:** 100 mph.
Seating Layout: 1: 2+2 facing, 2: 3+2 facing.
Multiple Working: Within class & with Classes 317, 318, 319, 320, 322 and 323.

Class 321/3.

DTCO. Lot No. 31053 1988–90. 16/57 (321 347–366 16/56). 29.7 t.
MSO. Lot No. 31054 1988–90. –/82. 51.5 t.
TSO. Lot No. 31055 1988–90. –/75 2T. 29.1 t.
DTSO. Lot No. 31056 1988–90. –/78. 29.7 t.

321301	**NX**	E	*GA*	IL	78049	62975	71880	77853
321302	**NX**	E	*GA*	IL	78050	62976	71881	77854
321303	**NX**	E	*GA*	IL	78051	62977	71882	77855
321304	**NX**	E	*GA*	IL	78052	62978	71883	77856
321305	**NX**	E	*GA*	IL	78053	62979	71884	77857
321306	**NX**	E	*GA*	IL	78054	62980	71885	77858
321307	**NX**	E	*GA*	IL	78055	62981	71886	77859
321308	**NX**	E	*GA*	IL	78056	62982	71887	77860

321 309	**NX**	E	*GA*	IL	78057	62983	71888	77861
321 310	**NX**	E	*GA*	IL	78058	62984	71889	77862
321 311	**NX**	E	*GA*	IL	78059	62985	71890	77863
321 312	**NX**	E	*GA*	IL	78060	62986	71891	77864
321 313	**NX**	E	*GA*	IL	78061	62987	71892	77865
321 314	**NX**	E	*GA*	IL	78062	62988	71893	77866
321 315	**NX**	E	*GA*	IL	78063	62989	71894	77867
321 316	**NX**	E	*GA*	IL	78064	62990	71895	77868
321 317	**NX**	E	*GA*	IL	78065	62991	71896	77869
321 318	**NX**	E	*GA*	IL	78066	62992	71897	77870
321 319	**NX**	E	*GA*	IL	78067	62993	71898	77871
321 320	**NX**	E	*GA*	IL	78068	62994	71899	77872
321 321	**NX**	E	*GA*	IL	78069	62995	71900	77873
321 322	**NX**	E	*GA*	IL	78070	62996	71901	77874
321 323	**NX**	E	*GA*	IL	78071	62997	71902	77875
321 324	**NX**	E	*GA*	IL	78072	62998	71903	77876
321 325	**NX**	E	*GA*	IL	78073	62999	71904	77877
321 326	**NX**	E	*GA*	IL	78074	63000	71905	77878
321 327	**NC**	E	*GA*	IL	78075	63001	71906	77879
321 328	**NX**	E	*GA*	IL	78076	63002	71907	77880
321 329	**NX**	E	*GA*	IL	78077	63003	71908	77881
321 330	**NC**	E	*GA*	IL	78078	63004	71909	77882
321 331	**NC**	E	*GA*	IL	78079	63005	71910	77883
321 332	**NC**	E	*GA*	IL	78080	63006	71911	77884
321 333	**NC**	E	*GA*	IL	78081	63007	71912	77885
321 334	**NC**	E	*GA*	IL	78082	63008	71913	77886
321 335	**NC**	E	*GA*	IL	78083	63009	71914	77887
321 336	**NC**	E	*GA*	IL	78084	63010	71915	77888
321 337	**NC**	E	*GA*	IL	78085	63011	71916	77889
321 338	**NC**	E	*GA*	IL	78086	63012	71917	77890
321 339	**NC**	E	*GA*	IL	78087	63013	71918	77891
321 340	**NC**	E	*GA*	IL	78088	63014	71919	77892
321 341	**NC**	E	*GA*	IL	78089	63015	71920	77893
321 342	**NC**	E	*GA*	IL	78090	63016	71921	77894
321 343	**NC**	E	*GA*	IL	78091	63017	71922	77895
321 344	**NC**	E	*GA*	IL	78092	63018	71923	77896
321 345	**NC**	E	*GA*	IL	78093	63019	71924	77897
321 346	**NC**	E	*GA*	IL	78094	63020	71925	77898
321 347	**NC**	E	*GA*	IL	78131	63105	71991	78280
321 348	**NC**	E	*GA*	IL	78132	63106	71992	78281
321 349	**NC**	E	*GA*	IL	78133	63107	71993	78282
321 350	**NC**	E	*GA*	IL	78134	63108	71994	78283
321 351	**NC**	E	*GA*	IL	78135	63109	71995	78284
321 352	**NC**	E	*GA*	IL	78136	63110	71996	78285
321 353	**NC**	E	*GA*	IL	78137	63111	71997	78286
321 354	**NC**	E	*GA*	IL	78138	63112	71998	78287
321 355	**NC**	E	*GA*	IL	78139	63113	71999	78288
321 356	**NC**	E	*GA*	IL	78140	63114	72000	78289
321 357	**NC**	E	*GA*	IL	78141	63115	72001	78290
321 358	**NC**	E	*GA*	IL	78142	63116	72002	78291
321 359	**GA**	E	*GA*	IL	78143	63117	72003	78292

321360	**NC**	E	*GA*	IL	78144	63118	72004	78293
321361	**GA**	E	*GA*	IL	78145	63119	72005	78294
321362	**GA**	E	*GA*	IL	78146	63120	72006	78295
321363	**GA**	E	*GA*	IL	78147	63121	72007	78296
321364	**GA**	E	*GA*	IL	78148	63122	72008	78297
321365	**GA**	E	*GA*	IL	78149	63123	72009	78298
321366	**GA**	E	*GA*	IL	78150	63124	72010	78299

Names (carried on TSO):

321312	Southend-on-Sea
321313	University of Essex
321321	NSPCC ESSEX FULL STOP
321334	Amsterdam
321336	GEOFFREY FREEMAN ALLEN
321342	R. Barnes
321343	RSA RAILWAY STUDY ASSOCIATION
321351	London Southend Airport
321361	Phoenix

Class 321/4.

The original vehicles 71966 and 77960 from 321418 and 78114 and 63082 from 321420 were written off after the Watford Junction accident in 1996. The undamaged vehicles were formed together as 321418 whilst four new vehicles were built in 1997, taking the same numbers as the scrapped vehicles, and these became the second 321420.

The DTCOs of 321421–437 have had 12 First Class seats declassified.

* 321448 has received an interior refurbishment as an Eversholt demonstrator unit. It has been fitted with two different types of interior using seats supplied by ATD. 78130 and 63104 have a "suburban" interior with a 3+2 seating layout and 78279 and 71990 have a "metro" interior with 2+2 seating. The unit is in normal service with Greater Anglia.

Non-standard livery: 321448 Eversholt demonstrator (silver with blue doors and multi-coloured stripes).

DTCO. Lot No. 31067 1989–90. 28/40 (321 421–437 16/52, 321 438–447 16/56). 29.8 t. (* 16/30(4) 1TD 2W 33.9 t).
MSO. Lot No. 31068 1989–90. –/79 (321 438–447 –/82). 51.6 t (* –/82. 54.0 t).
TSO. Lot No. 31069 1989–90. –/74 2T (321 438–447 –/75 2T). 29.2 t (* –/62 1T. 31.7 t).
DTSO. Lot No. 31070 1989–90. –/78. 29.8 t. (* –/58. 33.2 t).

321401	**FU**	E	*GT*	HE	78095	63063	71949	77943
321402	**FU**	E	*GT*	HE	78096	63064	71950	77944
321403	**FU**	E	*GT*	HE	78097	63065	71951	77945
321404	**FU**	E	*GT*	HE	78098	63066	71952	77946
321405	**FU**	E	*GT*	HE	78099	63067	71953	77947
321406	**FU**	E	*GT*	HE	78100	63068	71954	77948
321407	**FU**	E	*GT*	HE	78101	63069	71955	77949
321408	**FU**	E	*GT*	HE	78102	63070	71956	77950
321409	**FU**	E	*GT*	HE	78103	63071	71957	77951
321410	**FU**	E	*GT*	HE	78104	63072	71958	77952
321411	**LM**	E	*LM*	NN	78105	63073	71959	77953

321 412	**LM**	E	*LM*	NN	78106	63074	71960	77954
321 413	**LM**	E	*LM*	NN	78107	63075	71961	77955
321 414	**LM**	E	*LM*	NN	78108	63076	71962	77956
321 415	**LM**	E	*LM*	NN	78109	63077	71963	77957
321 416	**LM**	E	*LM*	NN	78110	63078	71964	77958
321 417	**LM**	E	*LM*	NN	78111	63079	71965	77959
321 418	**FU**	E	*GT*	HE	78112	63080	71968	77962
321 419	**FU**	E	*GT*	HE	78113	63081	71967	77961
321 420	**FU**	E	*GT*	HE	78114	63082	71966	77960
321 421	**NC**	E	*GA*	IL	78115	63083	71969	77963
321 422	**NC**	E	*GA*	IL	78116	63084	71970	77964
321 423	**NC**	E	*GA*	IL	78117	63085	71971	77965
321 424	**NX**	E	*GA*	IL	78118	63086	71972	77966
321 425	**NC**	E	*GA*	IL	78119	63087	71973	77967
321 426	**NX**	E	*GA*	IL	78120	63088	71974	77968
321 427	**NX**	E	*GA*	IL	78121	63089	71975	77969
321 428	**NX**	E	*GA*	IL	78122	63090	71976	77970
321 429	**NX**	E	*GA*	IL	78123	63091	71977	77971
321 430	**NX**	E	*GA*	IL	78124	63092	71978	77972
321 431	**NX**	E	*GA*	IL	78151	63125	72011	78300
321 432	**NC**	E	*GA*	IL	78152	63126	72012	78301
321 433	**NC**	E	*GA*	IL	78153	63127	72013	78302
321 434	**NC**	E	*GA*	IL	78154	63128	72014	78303
321 435	**NC**	E	*GA*	IL	78155	63129	72015	78304
321 436	**NC**	E	*GA*	IL	78156	63130	72016	78305
321 437	**NC**	E	*GA*	IL	78157	63131	72017	78306
321 438	**GA**	E	*GA*	IL	78158	63132	72018	78307
321 439	**GA**	E	*GA*	IL	78159	63133	72019	78308
321 440	**GA**	E	*GA*	IL	78160	63134	72020	78309
321 441	**GA**	E	*GA*	IL	78161	63135	72021	78310
321 442	**GA**	E	*GA*	IL	78162	63136	72022	78311
321 443	**GA**	E	*GA*	IL	78125	63099	71985	78274
321 444	**NC**	E	*GA*	IL	78126	63100	71986	78275
321 445	**NC**	E	*GA*	IL	78127	63101	71987	78276
321 446	**NC**	E	*GA*	IL	78128	63102	71988	78277
321 447	**GA**	E	*GA*	IL	78129	63103	71989	78278
321 448	* **0**	E	*GA*	IL	78130	63104	71990	78279

Names (carried on TSO):

321 403 Stewart Fleming Signalman King's Cross
321 409 Dame Alice Owen's School 400 Years of Learning
321 428 The Essex Commuter
321 442 Crouch Valley 1889–2014
321 444 Essex Lifeboats
321 446 George Mullings

Class 321/9. DTSO(A)–MSO–TSO–DTSO(B).

DTSO(A). Lot No. 31108 1991. –/70(8). 29.2 t.
MSO. Lot No. 31109 1991. –/79. 51.1 t.
TSO. Lot No. 31110 1991. –/74 2T. 29.0 t.
DTSO(B). Lot No. 31111 1991. –/70(7) 1W. 29.2 t.

321901	**YR**	E	*NO*	NL	77990	63153	72128	77993
321902	**YR**	E	*NO*	NL	77991	63154	72129	77994
321903	**YR**	E	*NO*	NL	77992	63155	72130	77995

CLASS 322   BREL YORK

Units built for use on Stansted Airport services, used for a number of years with ScotRail before transfer to Northern. Currently being refurbished and fitted with universal access toilets to comply with the 2020 accessibility regulations.

Formation: DTSO–MSO–TSO–DTSO.
Construction: Steel.
Traction Motors: Four Brush TM2141C of 268 kW.
Wheel Arrangement: 2-2 + Bo-Bo + 2-2 + 2-2.

Braking: Disc.	**Dimensions:** 19.95/19.92 x 2.82 m.
Bogies: P7-4 (MSO), T3-7 (others).	**Couplers:** Tightlock.
Gangways: Within unit.	**Control System:** Thyristor.
Doors: Sliding.	**Maximum Speed:** 100 mph.

Seating Layout: 3+2 facing.
Multiple Working: Within class & with Classes 317, 318, 319, 320, 321 and 323.

DTSO(A). Lot No. 31094 1990. –/73 (* –/54(4) 1TD 2W). 29.3 t (* 31.7 t).
MSO. Lot No. 31092 1990. –/83. 51.5 t (* 52.1 t).
TSO. Lot No. 31093 1990. –/76 2T (* –/80 1T). 28.8 t (* 30.6 t).
DTSO(B). Lot No. 31091 1990. –/71(3) 1W (* –/79). 29.1 t (*30.6 t).

322481	*	**NB**	E	*NO*	NL	78163	63137	72023	77985
322482	*	**NB**	E	*NO*	NL	78164	63138	72024	77986
322483	*	**NB**	E	*NO*	NL	78165	63139	72025	77987
322484		**FB**	E	*NO*	NL	78166	63140	72026	77988
322485		**FB**	E	*NO*	NL	78167	63141	72027	77989

CLASS 323 HUNSLET TRANSPORTATION PROJECTS

Suburban units.

Formation: DMSO–PTSO–DMSO.
Construction: Welded aluminium alloy.
Traction Motors: Four Holec DMKT 52/24 asynchronous of 146 kW.
Wheel Arrangement: Bo-Bo + 2-2 + Bo-Bo.

Braking: Disc.	**Dimensions:** 23.37/23.44 x 2.80 m.
Bogies: SRP BP62 (DMSO), BT52 (PTSO).	**Couplers:** Tightlock.
Gangways: Within unit.	**Control System:** GTO Inverter.
Doors: Sliding plug.	**Maximum Speed:** 90 mph.

Seating Layout: 3+2 facing/unidirectional.
Multiple Working: Within class & with Classes 317, 318, 319, 320, 321 and 322.

DMSO(A). Lot No. 31112 Hunslet 1992–93. –/98 (* –/82). 41.0 t.
TSO. Lot No. 31113 Hunslet 1992–93. –/88(5) 1T 2W. (* –/80 1T 2W). 39.4t.
DMSO(B). Lot No. 31114 Hunslet 1992–93. –/98 (* –/82). 41.0 t.

323 201		**LM**	P	*LM*	SO	64001	72201	65001
323 202		**LM**	P	*LM*	SO	64002	72202	65002
323 203		**LM**	P	*LM*	SO	64003	72203	65003
323 204		**LM**	P	*LM*	SO	64004	72204	65004
323 205		**LM**	P	*LM*	SO	64005	72205	65005
323 206		**LM**	P	*LM*	SO	64006	72206	65006
323 207		**LM**	P	*LM*	SO	64007	72207	65007
323 208		**LM**	P	*LM*	SO	64008	72208	65008
323 209		**LM**	P	*LM*	SO	64009	72209	65009
323 210		**LM**	P	*LM*	SO	64010	72210	65010
323 211		**LM**	P	*LM*	SO	64011	72211	65011
323 212		**LM**	P	*LM*	SO	64012	72212	65012
323 213		**LM**	P	*LM*	SO	64013	72213	65013
323 214		**LM**	P	*LM*	SO	64014	72214	65014
323 215		**LM**	P	*LM*	SO	64015	72215	65015
323 216		**LM**	P	*LM*	SO	64016	72216	65016
323 217		**LM**	P	*LM*	SO	64017	72217	65017
323 218		**LM**	P	*LM*	SO	64018	72218	65018
323 219		**LM**	P	*LM*	SO	64019	72219	65019
323 220		**LM**	P	*LM*	SO	64020	72220	65020
323 221		**LM**	P	*LM*	SO	64021	72221	65021
323 222		**LM**	P	*LM*	SO	64022	72222	65022
323 223	*	**NO**	P	*NO*	LG	64023	72223	65023
323 224	*	**NO**	P	*NO*	LG	64024	72224	65024
323 225	*	**NO**	P	*NO*	LG	64025	72225	65025
323 226		**NO**	P	*NO*	LG	64026	72226	65026
323 227		**NO**	P	*NO*	LG	64027	72227	65027
323 228		**NO**	P	*NO*	LG	64028	72228	65028
323 229		**NO**	P	*NO*	LG	64029	72229	65029
323 230		**NO**	P	*NO*	LG	64030	72230	65030
323 231		**NO**	P	*NO*	LG	64031	72231	65031
323 232		**NO**	P	*NO*	LG	64032	72232	65032
323 233		**NO**	P	*NO*	LG	64033	72233	65033
323 234		**NO**	P	*NO*	LG	64034	72234	65034
323 235		**NO**	P	*NO*	LG	64035	72235	65035
323 236		**NO**	P	*NO*	LG	64036	72236	65036
323 237		**NO**	P	*NO*	LG	64037	72237	65037
323 238		**NO**	P	*NO*	LG	64038	72238	65038
323 239		**NO**	P	*NO*	LG	64039	72239	65039
323 240		**LM**	P	*LM*	SO	64040	72340	65040
323 241		**LM**	P	*LM*	SO	64041	72341	65041
323 242		**LM**	P	*LM*	SO	64042	72342	65042
323 243		**LM**	P	*LM*	SO	64043	72343	65043

CLASS 325   ABB DERBY

Postal units based on Class 319s. Compatible with diesel or electric locomotive haulage.

Formation: DTPMV–MPMV–TPMV–DTPMV.
System: 25 kV AC overhead/750 V DC third rail.
Construction: Steel.
Traction Motors: Four GEC G315BZ of 268 kW.
Wheel Arrangement: 2-2 + Bo-Bo + 2-2 + 2-2.
Braking: Disc.   **Dimensions:** 19.33 x 2.82 m.
Bogies: P7-4 (MSO), T3-7 (others).   **Couplers:** Drop-head buckeye.
Gangways: None.   **Control System:** GTO Chopper.
Doors: Roller shutter.   **Maximum Speed:** 100 mph.
Multiple Working: Within class.

DTPMV. Lot No. 31144 1995. 29.1 t.
MPMV. Lot No. 31145 1995. 49.5 t.
TPMV. Lot No. 31146 1995. 30.7 t.

325 001	**RM**	RM	*DB*	CE	68300	68340	68360	68301
325 002	**RM**	RM	*DB*	CE	68302	68341	68361	68303
325 003	**RM**	RM	*DB*	CE	68304	68342	68362	68305
325 004	**RM**	RM	*DB*	CE	68306	68343	68363	68307
325 005	**RM**	RM	*DB*	CE	68308	68344	68364	68309
325 006	**RM**	RM	*DB*	CE	68310	68345	68365	68311
325 007	**RM**	RM	*DB*	CE	68312	68346	68366	68313
325 008	**RM**	RM	*DB*	CE	68314	68347	68367	68315
325 009	**RM**	RM	*DB*	CE	68316	68349	68368	68317
325 011	**RM**	RM	*DB*	CE	68320	68350	68370	68321
325 012	**RM**	RM	*DB*	CE	68322	68351	68371	68323
325 013	**RM**	RM	*DB*	CE	68324	68352	68372	68325
325 014	**RM**	RM	*DB*	CE	68326	68353	68373	68327
325 015	**RM**	RM	*DB*	CE	68328	68354	68374	68329
325 016	**RM**	RM	*DB*	CE	68330	68355	68375	68331

Names (carried on one side of each DTPMV):

325 002 Royal Mail North Wales & North West
325 006 John Grierson
325 008 Peter Howarth CBE

CLASS 332   HEATHROW EXPRESS CAF/SIEMENS

Dedicated Heathrow Express units. Five units were increased from 4-car to 5-car in 2002. Usually operate in coupled pairs.

Formations: DMSO–TSO–PTSO–(TSO)–DMFO.
Construction: Steel.
Traction Motors: Two Siemens monomotors asynchronous of 350 kW.
Wheel Arrangement: B-B + 2-2 + 2-2 (+ 2-2) + B-B.
Braking: Disc.   **Dimensions:** 23.74/23.35/23.14 x 2.75 m.
Bogies: CAF.   **Couplers:** Scharfenberg 10L.
Gangways: Within unit.   **Control System:** IGBT Inverter.
Doors: Sliding plug.   **Maximum Speed:** 100 mph.
Heating & ventilation: Air conditioning.
Seating: 1: 1+1 facing/unidirectional, 2: 2+2 mainly unidirectional.
Multiple Working: Within class.

DMSO. CAF 1997–98. –/43 (8). 49.9 t.
72400–413. TSO. CAF 1997–98. –/64 (11). 38.4 t.
72414–418. TSO. CAF 2002. –/56 35.8 t.
PTSO. CAF 1997–98. –/39 (11) 1TD 2W. 47.6 t.
DMFO. CAF 1997–98. 20/–. 49.5 t.

332 001	**HE**	HE	*HE*	OH	78400	72412	63400		78401
332 002	**HE**	HE	*HE*	OH	78402	72409	63406		78403
332 003	**HE**	HE	*HE*	OH	78404	72407	63402		78405
332 004	**HE**	HE	*HE*	OH	78406	72405	63403		78407
332 005	**HE**	HE	*HE*	OH	78408	72411	63404	72417	78409
332 006	**HE**	HE	*HE*	OH	78410	72410	63405	72415	78411
332 007	**HE**	HE	*HE*	OH	78412	72401	63401	72414	78413
332 008	**HE**	HE	*HE*	OH	78414	72413	63407	72418	78415
332 009	**HE**	HE	*HE*	OH	78416	72400	63408	72416	78417
332 010	**HE**	HE	*HE*	OH	78418	72402	63409		78419
332 011	**HE**	HE	*HE*	OH	78420	72403	63410		78421
332 012	**HE**	HE	*HE*	OH	78422	72404	63411		78423
332 013	**HE**	HE	*HE*	OH	78424	72408	63412		78425
332 014	**HE**	HE	*HE*	OH	78426	72406	63413		78427

CLASS 333   CAF/SIEMENS

West Yorkshire area suburban units.

Formation: DMSO–PTSO–TSO–DMSO.
Construction: Steel.
Traction Motors: Two Siemens monomotors asynchronous of 350 kW.
Wheel Arrangement: B-B + 2-2 + 2-2 + B-B.
Braking: Disc.
Dimensions: 23.74 (outer ends)/23.35 (TSO) x 2.75 m.
Bogies: CAF.   **Couplers:** Dellner 10L.
Gangways: Within unit.   **Control System:** IGBT Inverter.
Doors: Sliding plug.   **Maximum Speed:** 100 mph.
Heating & ventilation: Air conditioning.

Seating Layout: 3+2 facing/unidirectional.
Multiple Working: Within class.

333001–008 were made up to 4-car units from 3-car units in 2002.

333009–016 were made up to 4-car units from 3-car units in 2003.

DMSO(A). (Odd Nos.) CAF 2001. –/90. 50.6 t.
PTSO. CAF 2001. –/73(6) 1TD 2W. 46.0 t.
TSO. CAF 2002–03. –/100. 38.5 t.
DMSO(B). (Even Nos.) CAF 2001. –/90. 50.0 t.

333001	**YR**	A	*NO*	NL	78451	74461	74477	78452
333002	**YR**	A	*NO*	NL	78453	74462	74478	78454
333003	**YR**	A	*NO*	NL	78455	74463	74479	78456
333004	**YR**	A	*NO*	NL	78457	74464	74480	78458
333005	**YR**	A	*NO*	NL	78459	74465	74481	78460
333006	**YR**	A	*NO*	NL	78461	74466	74482	78462
333007	**YR**	A	*NO*	NL	78463	74467	74483	78464
333008	**YR**	A	*NO*	NL	78465	74468	74484	78466
333009	**YR**	A	*NO*	NL	78467	74469	74485	78468
333010	**YR**	A	*NO*	NL	78469	74470	74486	78470
333011	**YR**	A	*NO*	NL	78471	74471	74487	78472
333012	**YR**	A	*NO*	NL	78473	74472	74488	78474
333013	**YR**	A	*NO*	NL	78475	74473	74489	78476
333014	**YR**	A	*NO*	NL	78477	74474	74490	78478
333015	**YR**	A	*NO*	NL	78479	74475	74491	78480
333016	**YR**	A	*NO*	NL	78481	74476	74492	78482

Name (carried on end cars):

33᠎ 007 Alderman J Arthur Godwin First Lord Mayor of Bradford 1907

CLASS 334   JUNIPER   ALSTOM BIRMINGHAM

Outer suburban units.

Formation: DMSO–PTSO–DMSO.
Construction: Steel.
Traction Motors: Two Alstom ONIX 800 asynchronous of 270 kW.
Wheel Arrangement: 2-Bo + 2-2 + Bo-2.
Braking: Disc.   **Dimensions:** 21.01/19.94 x 2.80 m.
Bogies: Alstom LTB3/TBP3.   **Couplers:** Tightlock.
Gangways: Within unit.   **Control System:** IGBT Inverter.
Doors: Sliding plug.   **Maximum Speed:** 90 mph.
Heating & ventilation: Pressure heating and ventilation.
Seating Layout: 2+2 facing/unidirectional (3+2 in PTSO).
Multiple Working: Within class.

DMSO(A). Alstom Birmingham 1999–2001. –/64. 42.6 t.
PTSO. Alstom Birmingham 1999–2001. –/55 1TD 1W. 39.4 t.
DMSO(B). Alstom Birmingham 1999–2001. –/64. 42.6 t.

334001	**SR**	E	*SR*	GW	64101	74301	65101
334002	**SR**	E	*SR*	GW	64102	74302	65102
334003	**SR**	E	*SR*	GW	64103	74303	65103
334004	**SR**	E	*SR*	GW	64104	74304	65104
334005	**SR**	E	*SR*	GW	64105	74305	65105
334006	**SR**	E	*SR*	GW	64106	74306	65106
334007	**SR**	E	*SR*	GW	64107	74307	65107
334008	**SR**	E	*SR*	GW	64108	74308	65108
334009	**SR**	E	*SR*	GW	64109	74309	65109
334010	**SR**	E	*SR*	GW	64110	74310	65110
334011	**SR**	E	*SR*	GW	64111	74311	65111
334012	**SR**	E	*SR*	GW	64112	74312	65112
334013	**SR**	E	*SR*	GW	64113	74313	65113
334014	**SR**	E	*SR*	GW	64114	74314	65114
334015	**SR**	E	*SR*	GW	64115	74315	65115
334016	**SR**	E	*SR*	GW	64116	74316	65116
334017	**SR**	E	*SR*	GW	64117	74317	65117
334018	**SR**	E	*SR*	GW	64118	74318	65118
334019	**SR**	E	*SR*	GW	64119	74319	65119
334020	**SR**	E	*SR*	GW	64120	74320	65120
334021	**SR**	E	*SR*	GW	64121	74321	65121
334022	**SR**	E	*SR*	GW	64122	74322	65122
334023	**SR**	E	*SR*	GW	64123	74323	65123
334024	**SR**	E	*SR*	GW	64124	74324	65124
334025	**SR**	E	*SR*	GW	64125	74325	65125
334026	**SR**	E	*SR*	GW	64126	74326	65126
334027	**SR**	E	*SR*	GW	64127	74327	65127
334028	**SR**	E	*SR*	GW	64128	74328	65128
334029	**SR**	E	*SR*	GW	64129	74329	65129
334030	**SR**	E	*SR*	GW	64130	74330	65130
334031	**SR**	E	*SR*	GW	64131	74331	65131
334032	**SR**	E	*SR*	GW	64132	74332	65132
334033	**SR**	E	*SR*	GW	64133	74333	65133
334034	**SR**	E	*SR*	GW	64134	74334	65134
334035	**SR**	E	*SR*	GW	64135	74335	65135
334036	**SR**	E	*SR*	GW	64136	74336	65136
334037	**SR**	E	*SR*	GW	64137	74337	65137
334038	**SR**	E	*SR*	GW	64138	74338	65138
334039	**SR**	E	*SR*	GW	64139	74339	65139
334040	**SR**	E	*SR*	GW	64140	74340	65140

CLASS 350   DESIRO UK   SIEMENS

Outer suburban and long distance units.

Formation: DMCO–TCO–PTSO–DMCO.
Systems: 25 kV AC overhead (350/1s built with 750 V DC).
Construction: Welded aluminium.
Traction Motors: 4 Siemens 1TB2016-0GB02 asynchronous of 250 kW.
Wheel Arrangement: Bo-Bo + 2-2 + 2-2 + Bo-Bo.
Braking: Disc & regenerative.   **Dimensions:** 20.34 x 2.79 m.
Bogies: SGP SF5000.   **Couplers:** Dellner 12.
Gangways: Throughout.   **Control System:** IGBT Inverter.
Doors: Sliding plug.
Maximum Speed: 110 mph (350/1, 350/3 & 350/4) and 100 mph (350/2).
Heating & ventilation: Air conditioning.
Seating Layout: Various, see sub-class headings.
Multiple Working: Within class.

Class 350/1. Original build units owned by Angel Trains. Formerly part of an aborted South West Trains 5-car Class 450/2 order. 2+2 seating.

Seating Layout: 1: 2+2 facing, 2: 2+2 facing/unidirectional.

Advertising livery: 350 110 Project 110 (silver centre cars).

DMSO(A). Siemens Krefeld 2004–05. –/60. 48.7 t.
TCO. Siemens Krefeld/Prague 2004–05. 24/32 1T. 36.2 t.
PTSO. Siemens Krefeld/Prague 2004–05. –/50(9) 1TD 2W. 45.2 t.
DMSO(B). Siemens Krefeld 2004–05. –/60. 49.2 t.

350 101	**LM**	A	*LM*	NN	63761	66811	66861	63711
350 102	**LM**	A	*LM*	NN	63762	66812	66862	63712
350 103	**LM**	A	*LM*	NN	63765	66813	66863	63713
350 104	**LM**	A	*LM*	NN	63764	66814	66864	63714
350 105	**LM**	A	*LM*	NN	63763	66815	66868	63715
350 106	**LM**	A	*LM*	NN	63766	66816	66866	63716
350 107	**LM**	A	*LM*	NN	63767	66817	66867	63717
350 108	**LM**	A	*LM*	NN	63768	66818	66865	63718
350 109	**LM**	A	*LM*	NN	63769	66819	66869	63719
350 110	**AL**	A	*LM*	NN	63770	66820	66870	63720
350 111	**LM**	A	*LM*	NN	63771	66821	66871	63721
350 112	**LM**	A	*LM*	NN	63772	66822	66872	63722
350 113	**LM**	A	*LM*	NN	63773	66823	66873	63723
350 114	**LM**	A	*LM*	NN	63774	66824	66874	63724
350 115	**LM**	A	*LM*	NN	63775	66825	66875	63725
350 116	**LM**	A	*LM*	NN	63776	66826	66876	63726
350 117	**LM**	A	*LM*	NN	63777	66827	66877	63727
350 118	**LM**	A	*LM*	NN	63778	66828	66878	63728
350 119	**LM**	A	*LM*	NN	63779	66829	66879	63729
350 120	**LM**	A	*LM*	NN	63780	66830	66880	63730
350 121	**LM**	A	*LM*	NN	63781	66831	66881	63731
350 122	**LM**	A	*LM*	NN	63782	66832	66882	63732
350 123	**LM**	A	*LM*	NN	63783	66833	66883	63733
350 124	**LM**	A	*LM*	NN	63784	66834	66884	63734

350 125	**LM**	A	*LM*	NN	63785	66835	66885	63735
350 126	**LM**	A	*LM*	NN	63786	66836	66886	63736
350 127	**LM**	A	*LM*	NN	63787	66837	66887	63737
350 128	**LM**	A	*LM*	NN	63788	66838	66888	63738
350 129	**LM**	A	*LM*	NN	63789	66839	66889	63739
350 130	**LM**	A	*LM*	NN	63790	66840	66890	63740

Class 350/2. Owned by Porterbrook Leasing.

Seating Layout: 1: 2+2 facing, 2: 3+2 facing/unidirectional.

DMSO(A). Siemens Krefeld 2008–09. –/70. 43.7 t.
TCO. Siemens Prague 2008–09. 24/42 1T. 35.3 t.
PTSO. Siemens Prague 2008–09. –/61(9) 1TD 2W. 42.9 t.
DMSO(B). Siemens Krefeld 2008–09. –/70. 44.2 t.

350 231	**LM**	P	*LM*	NN	61431	65231	67531	61531
350 232	**LM**	P	*LM*	NN	61432	65232	67532	61532
350 233	**LM**	P	*LM*	NN	61433	65233	67533	61533
350 234	**LM**	P	*LM*	NN	61434	65234	67534	61534
350 235	**LM**	P	*LM*	NN	61435	65235	67535	61535
350 236	**LM**	P	*LM*	NN	61436	65236	67536	61536
350 237	**LM**	P	*LM*	NN	61437	65237	67537	61537
350 238	**LM**	P	*LM*	NN	61438	65238	67538	61538
350 239	**LM**	P	*LM*	NN	61439	65239	67539	61539
350 240	**LM**	P	*LM*	NN	61440	65240	67540	61540
350 241	**LM**	P	*LM*	NN	61441	65241	67541	61541
350 242	**LM**	P	*LM*	NN	61442	65242	67542	61542
350 243	**LM**	P	*LM*	NN	61443	65243	67543	61543
350 244	**LM**	P	*LM*	NN	61444	65244	67544	61544
350 245	**LM**	P	*LM*	NN	61445	65245	67545	61545
350 246	**LM**	P	*LM*	NN	61446	65246	67546	61546
350 247	**LM**	P	*LM*	NN	61447	65247	67547	61547
350 248	**LM**	P	*LM*	NN	61448	65248	67548	61548
350 249	**LM**	P	*LM*	NN	61449	65249	67549	61549
350 250	**LM**	P	*LM*	NN	61450	65250	67550	61550
350 251	**LM**	P	*LM*	NN	61451	65251	67551	61551
350 252	**LM**	P	*LM*	NN	61452	65252	67552	61552
350 253	**LM**	P	*LM*	NN	61453	65253	67553	61553
350 254	**LM**	P	*LM*	NN	61454	65254	67554	61554
350 255	**LM**	P	*LM*	NN	61455	65255	67555	61555
350 256	**LM**	P	*LM*	NN	61456	65256	67556	61556
350 257	**LM**	P	*LM*	NN	61457	65257	67557	61557
350 258	**LM**	P	*LM*	NN	61458	65258	67558	61558
350 259	**LM**	P	*LM*	NN	61459	65259	67559	61559
350 260	**LM**	P	*LM*	NN	61460	65260	67560	61560
350 261	**LM**	P	*LM*	NN	61461	65261	67561	61561
350 262	**LM**	P	*LM*	NN	61462	65262	67562	61562
350 263	**LM**	P	*LM*	NN	61463	65263	67563	61563
350 264	**LM**	P	*LM*	NN	61464	65264	67564	61564
350 265	**LM**	P	*LM*	NN	61465	65265	67565	61565
350 266	**LM**	P	*LM*	NN	61466	65266	67566	61566
350 267	**LM**	P	*LM*	NN	61467	65267	67567	61567

Name (carried on one side of PTSO): 350 232   Chad Varah

Class 350/3. Owned by Angel Trains. London Midland units built for 110 mph operation.

Seating Layout: 1: 2+2 facing, 2: 2+2 facing/unidirectional.

DMSO(A). Siemens Krefeld 2014. –/60. 44.2 t.
TCO. Siemens Krefeld 2014. 24/36 1T. 36.3 t.
PTSO. Siemens Krefeld 2014. –/50(9) 1TD 2W. 44.0 t.
DMSO(B). Siemens Krefeld 2014. –/60. 45.0 t.

350368	**LM**	A	*LM*	NN	60141	60511	60651	60151
350369	**LM**	A	*LM*	NN	60142	60512	60652	60152
350370	**LM**	A	*LM*	NN	60143	60513	60653	60153
350371	**LM**	A	*LM*	NN	60144	60514	60654	60154
350372	**LM**	A	*LM*	NN	60145	60515	60655	60155
350373	**LM**	A	*LM*	NN	60146	60516	60656	60156
350374	**LM**	A	*LM*	NN	60147	60517	60657	60157
350375	**LM**	A	*LM*	NN	60148	60518	60658	60158
350376	**LM**	A	*LM*	NN	60149	60519	60659	60159
350377	**LM**	A	*LM*	NN	60150	60520	60660	60160

Class 350/4. Owned by Angel Trains. TransPennine Express units used on the Manchester Airport–Edinburgh/Glasgow route.

Seating Layout: 1: 2+1 facing, 2: 2+2 facing/unidirectional.

DMSO(A). Siemens Krefeld 2013–14. –/56. 44.2 t.
TCO. Siemens Krefeld 2013–14. 19/24 1T. 36.2 t.
PTSO. Siemens Krefeld 2013–14. –/42 1TD 1T. 44.6 t.
DMSO(B). Siemens Krefeld 2013–14. –/56. 45.0 t.

350401	**FT**	A	*TP*	AK	60691	60901	60941	60671
350402	**FT**	A	*TP*	AK	60692	60902	60942	60672
350403	**FT**	A	*TP*	AK	60693	60903	60943	60673
350404	**FT**	A	*TP*	AK	60694	60904	60944	60674
350405	**FT**	A	*TP*	AK	60695	60905	60945	60675
350406	**FT**	A	*TP*	AK	60696	60906	60946	60676
350407	**FT**	A	*TP*	AK	60697	60907	60947	60677
350408	**FT**	A	*TP*	AK	60698	60908	60948	60678
350409	**FT**	A	*TP*	AK	60699	60909	60949	60679
350410	**FT**	A	*TP*	AK	60700	60910	60950	60680

CLASS 357 ELECTROSTAR
ADTRANZ/BOMBARDIER DERBY

Provision for 750 V DC supply if required.

Formation: DMSO–MSO–PTSO–DMSO.
Construction: Welded aluminium alloy underframe, sides and roof with steel ends. All sections bolted together.
Traction Motors: Two Adtranz asynchronous of 250 kW.
Wheel Arrangement: 2-Bo + 2-Bo + 2-2 + Bo-2.
Braking: Disc & regenerative.　　　**Dimensions:** 20.40/19.99 x 2.80 m.
Bogies: Adtranz P3-25/T3-25.　　　**Couplers:** Tightlock.
Gangways: Within unit.　　　**Control System:** IGBT Inverter.
Doors: Sliding plug.　　　**Maximum Speed:** 100 mph.
Heating & ventilation: Air conditioning.
Seating Layout: 3+2 facing/unidirectional.
Multiple Working: Within class.

Class 357/0. Owned by Porterbrook Leasing.

DMSO(A). Adtranz Derby 1999–2001. –/71. 40.7 t.
MSO. Adtranz Derby 1999–2001. –/78. 36.7 t.
PTSO. Adtranz Derby 1999–2001. –/58(4) 1TD 2W. 39.5 t.
DMSO(B). Adtranz Derby 1999–2001. –/71. 40.7 t.

357001	**NC**	P	*C2*	EM	67651	74151	74051	67751
357002	**NC**	P	*C2*	EM	67652	74152	74052	67752
357003	**NC**	P	*C2*	EM	67653	74153	74053	67753
357004	**NC**	P	*C2*	EM	67654	74154	74054	67754
357005	**NC**	P	*C2*	EM	67655	74155	74055	67755
357006	**NC**	P	*C2*	EM	67656	74156	74056	67756
357007	**NC**	P	*C2*	EM	67657	74157	74057	67757
357008	**NC**	P	*C2*	EM	67658	74158	74058	67758
357009	**NC**	P	*C2*	EM	67659	74159	74059	67759
357010	**NC**	P	*C2*	EM	67660	74160	74060	67760
357011	**NC**	P	*C2*	EM	67661	74161	74061	67761
357012	**NC**	P	*C2*	EM	67662	74162	74062	67762
357013	**NC**	P	*C2*	EM	67663	74163	74063	67763
357014	**NC**	P	*C2*	EM	67664	74164	74064	67764
357015	**NC**	P	*C2*	EM	67665	74165	74065	67765
357016	**NC**	P	*C2*	EM	67666	74166	74066	67766
357017	**NC**	P	*C2*	EM	67667	74167	74067	67767
357018	**NC**	P	*C2*	EM	67668	74168	74068	67768
357019	**NC**	P	*C2*	EM	67669	74169	74069	67769
357020	**NC**	P	*C2*	EM	67670	74170	74070	67770
357021	**NC**	P	*C2*	EM	67671	74171	74071	67771
357022	**NC**	P	*C2*	EM	67672	74172	74072	67772
357023	**NC**	P	*C2*	EM	67673	74173	74073	67773
357024	**NC**	P	*C2*	EM	67674	74174	74074	67774
357025	**NC**	P	*C2*	EM	67675	74175	74075	67775
357026	**NC**	P	*C2*	EM	67676	74176	74076	67776

357027	**NC**	P	C2	EM	67677	74177	74077	67777
357028	**NC**	P	C2	EM	67678	74178	74078	67778
357029	**NC**	P	C2	EM	67679	74179	74079	67779
357030	**NC**	P	C2	EM	67680	74180	74080	67780
357031	**NC**	P	C2	EM	67681	74181	74081	67781
357032	**NC**	P	C2	EM	67682	74182	74082	67782
357033	**NC**	P	C2	EM	67683	74183	74083	67783
357034	**NC**	P	C2	EM	67684	74184	74084	67784
357035	**NC**	P	C2	EM	67685	74185	74085	67785
357036	**NC**	P	C2	EM	67686	74186	74086	67786
357037	**NC**	P	C2	EM	67687	74187	74087	67787
357038	**NC**	P	C2	EM	67688	74188	74088	67788
357039	**NC**	P	C2	EM	67689	74189	74089	67789
357040	**NC**	P	C2	EM	67690	74190	74090	67790
357041	**NC**	P	C2	EM	67691	74191	74091	67791
357042	**NC**	P	C2	EM	67692	74192	74092	67792
357043	**NC**	P	C2	EM	67693	74193	74093	67793
357044	**NC**	P	C2	EM	67694	74194	74094	67794
357045	**NC**	P	C2	EM	67695	74195	74095	67795
357046	**NC**	P	C2	EM	67696	74196	74096	67796

Names (carried on DMSO(A) and DMSO(B) (one plate on each)):

357001 BARRY FLAXMAN
357002 ARTHUR LEWIS STRIDE 1841–1922
357003 SOUTHEND city.on.sea
357004 TONY AMOS
357006 DIAMOND JUBILEE 1952–2012
357011 JOHN LOWING
357028 London, Tilbury & Southend Railway 1854–2004
357029 THOMAS WHITELEGG 1840–1922
357030 ROBERT HARBEN WHITELEGG 1871–1957

Class 357/2. Owned by Angel Trains.

DMSO(A). Bombardier Derby 2001–02. –/71. 40.7 t.
MSO. Bombardier Derby 2001–02. –/78. 36.7 t.
PTSO. Bombardier Derby 2001–02. –/58(4) 1TD 2W. 39.5 t.
DMSO(B). Bombardier Derby 2001–02. –/71. 40.7 t.

357201	**NC**	A	C2	EM	68601	74701	74601	68701
357202	**NC**	A	C2	EM	68602	74702	74602	68702
357203	**NC**	A	C2	EM	68603	74703	74603	68703
357204	**NC**	A	C2	EM	68604	74704	74604	68704
357205	**NC**	A	C2	EM	68605	74705	74605	68705
357206	**NC**	A	C2	EM	68606	74706	74606	68706
357207	**NC**	A	C2	EM	68607	74707	74607	68707
357208	**NC**	A	C2	EM	68608	74708	74608	68708
357209	**NC**	A	C2	EM	68609	74709	74609	68709
357210	**NC**	A	C2	EM	68610	74710	74610	68710
357211	**NC**	A	C2	EM	68611	74711	74611	68711
357212	**NC**	A	C2	EM	68612	74712	74612	68712
357213	**NC**	A	C2	EM	68613	74713	74613	68713

357 214	**NC**	A	*C2*	EM	68614	74714	74614	68714
357 215	**NC**	A	*C2*	EM	68615	74715	74615	68715
357 216	**NC**	A	*C2*	EM	68616	74716	74616	68716
357 217	**NC**	A	*C2*	EM	68617	74717	74617	68717
357 218	**NC**	A	*C2*	EM	68618	74718	74618	68718
357 219	**NC**	A	*C2*	EM	68619	74719	74619	68719
357 220	**NC**	A	*C2*	EM	68620	74720	74620	68720
357 221	**NC**	A	*C2*	EM	68621	74721	74621	68721
357 222	**NC**	A	*C2*	EM	68622	74722	74622	68722
357 223	**NC**	A	*C2*	EM	68623	74723	74623	68723
357 224	**NC**	A	*C2*	EM	68624	74724	74624	68724
357 225	**NC**	A	*C2*	EM	68625	74725	74625	68725
357 226	**NC**	A	*C2*	EM	68626	74726	74626	68726
357 227	**NC**	A	*C2*	EM	68627	74727	74627	68727
357 228	**NC**	A	*C2*	EM	68628	74728	74628	68728

Names (carried on DMSO(A) and DMSO(B) (one plate on each)):

357 201	KEN BIRD	357 207	JOHN PAGE
357 202	KENNY MITCHELL	357 208	DAVE DAVIS
357 203	HENRY PUMFRETT	357 209	JAMES SNELLING
357 204	DEREK FOWERS	357 213	UPMINSTER I.E.C.C.
357 205	JOHN D'SILVA	357 217	ALLAN BURNELL
357 206	MARTIN AUNGIER	357 227	SOUTHEND UNITED

CLASS 360/0   DESIRO UK   SIEMENS

Outer suburban/express units.

Formation: DMCO–PTSO–TSO–DMCO.
Construction: Welded aluminium.
Traction Motors: Four Siemens 1TB2016-0GB02 asynchronous of 250 kW.
Wheel Arrangement: Bo-Bo + 2-2 + 2-2 + Bo-Bo.
Braking: Disc & regenerative.   **Dimensions:** 20.34 x 2.80 m.
Bogies: SGP SF5000.   **Couplers:** Dellner 12.
Gangways: Within unit.   **Control System:** IGBT Inverter.
Doors: Sliding plug.   **Maximum Speed:** 100 mph.
Heating & ventilation: Air conditioning.
Seating Layout: 1: 2+2 facing, 2: 3+2 facing/unidirectional.
Multiple Working: Within class.

DMCO(A). Siemens Krefeld 2002–03. 8/59. 45.0 t.
PTSO. Siemens Vienna 2002–03. –/60(9) 1TD 2W. 43.0 t.
TSO. Siemens Vienna 2002–03. –/78. 35.0 t.
DMCO(B). Siemens Krefeld 2002–03. 8/59. 45.0 t.

360 101	**FB**	A	*GA*	IL	65551	72551	74551	68551
360 102	**FB**	A	*GA*	IL	65552	72552	74552	68552
360 103	**FB**	A	*GA*	IL	65553	72553	74553	68553
360 104	**FB**	A	*GA*	IL	65554	72554	74554	68554
360 105	**FB**	A	*GA*	IL	65555	72555	74555	68555
360 106	**FB**	A	*GA*	IL	65556	72556	74556	68556

360 107	**FB**	A	*GA*	IL	65557	72557	74557	68557
360 108	**FB**	A	*GA*	IL	65558	72558	74558	68558
360 109	**FB**	A	*GA*	IL	65559	72559	74559	68559
360 110	**FB**	A	*GA*	IL	65560	72560	74560	68560
360 111	**FB**	A	*GA*	IL	65561	72561	74561	68561
360 112	**FB**	A	*GA*	IL	65562	72562	74562	68562
360 113	**FB**	A	*GA*	IL	65563	72563	74563	68563
360 114	**FB**	A	*GA*	IL	65564	72564	74564	68564
360 115	**FB**	A	*GA*	IL	65565	72565	74565	68565
360 116	**FB**	A	*GA*	IL	65566	72566	74566	68566
360 117	**FB**	A	*GA*	IL	65567	72567	74567	68567
360 118	**FB**	A	*GA*	IL	65568	72568	74568	68568
360 119	**FB**	A	*GA*	IL	65569	72569	74569	68569
360 120	**FB**	A	*GA*	IL	65570	72570	74570	68570
360 121	**FB**	A	*GA*	IL	65571	72571	74571	68571

CLASS 360/2 DESIRO UK SIEMENS

4-car Class 350 testbed units rebuilt for use by Heathrow Express on Paddington–Heathrow Airport stopping services ("Heathrow Connect").

Original 4-car sets 360 201–204 were made up to 5-cars during 2007 using additional TSOs. A fifth unit (360 205) was delivered in late 2005 as a 5-car set. This set is normally used on Terminals 1&3–Terminal 4 shuttle services.

Formation: DMSO–PTSO–TSO–TSO–DMSO.
Construction: Welded aluminium.
Traction Motors: Four Siemens 1TB2016-0GB02 asynchronous of 250 kW.
Wheel Arrangement: Bo-Bo + 2-2 + 2-2 + 2-2 + Bo-Bo.
Braking: Disc & regenerative. **Dimensions:** 20.34 x 2.80 m.
Bogies: SGP SF5000. **Couplers:** Dellner 12.
Gangways: Within unit. **Control System:** IGBT Inverter.
Doors: Sliding plug. **Maximum Speed:** 100 mph.
Heating & ventilation: Air conditioning.
Seating Layout: 3+2 (* 2+2) facing/unidirectional.
Multiple Working: Within class.

DMSO(A). Siemens Krefeld 2002–06. –/63 (* –/54). 44.8 t.
PTSO. Siemens Krefeld 2002–06. –/57(9) 1TD 2W (* –/48(9) 2W). 44.2 t.
TSO. Siemens Krefeld 2005–06. –/74 (* –/62). 35.3 t.
TSO. Siemens Krefeld 2002–06. –/74 (* –/62). 34.1 t.
DMSO(B). Siemens Krefeld 2002–06. –/63 (* –/54). 44.4 t.

360 201		**HC**	HE	*HC*	OH	78431	63421	72431	72421	78441
360 202		**HC**	HE	*HC*	OH	78432	63422	72432	72422	78442
360 203		**HC**	HE	*HC*	OH	78433	63423	72433	72423	78443
360 204		**HC**	HE	*HC*	OH	78434	63424	72434	72424	78444
360 205	*	**HE**	HE	*HE*	OH	78435	63425	72435	72425	78445

CLASS 365   NETWORKER EXPRESS   ABB YORK

Outer suburban units.

Formations: DMCO–TSO–PTSO–DMCO.
Systems: 25 kV AC overhead but with 750 V DC third rail capability (units marked * were formerly used on DC lines in the South-East).
Construction: Welded aluminium alloy.
Traction Motors: Four GEC-Alsthom G354CX asynchronous of 157 kW.
Wheel Arrangement: Bo-Bo + 2-2 + 2-2 + Bo-Bo.
Braking: Disc & rheostatic.
Dimensions: 20.89/20.06 x 2.81 m.
Bogies: ABB P3-16/T3-16.   **Couplers:** Tightlock.
Gangways: Within unit.   **Control System:** GTO Inverter.
Doors: Sliding plug.   **Maximum Speed:** 100 mph.
Seating Layout: 1: 2+2 facing, 2: 2+2 facing.
Multiple Working: Within class only.

Advertising liveries:

365510   Cambridge & Ely; Cathedral cities (blue & white with various images).
365531   Nelson's County; Norfolk (blue & white with various images).

DMCO(A). Lot No. 31133 1994–95. 12/56. 41.7 t.
TSO. Lot No. 31134 1994–95. –/65 1TD (* –/64 1TD) 32.9 t.
PTSO. Lot No. 31135 1994–95. –/68 1T. 34.6 t.
DMCO(B). Lot No. 31136 1994–95. 12/56. 41.7 t.

365501	*	FU	E	*GT*	HE	65894	72241	72240	65935
365502	*	FU	E	*GT*	HE	65895	72243	72242	65936
365503	*	FU	E	*GT*	HE	65896	72245	72244	65937
365504	*	FU	E	*GT*	HE	65897	72247	72246	65938
365505	*	FU	E	*GT*	HE	65898	72249	72248	65939
365506	*	FU	E	*GT*	HE	65899	72251	72250	65940
365507	*	FU	E	*GT*	HE	65900	72253	72252	65941
365508	*	FU	E	*GT*	HE	65901	72255	72254	65942
365509	*	FU	E	*GT*	HE	65902	72257	72256	65943
365510	*	AL	E	*GT*	HE	65903	72259	72258	65944
365511	*	FU	E	*GT*	HE	65904	72261	72260	65945
365512	*	FU	E	*GT*	HE	65905	72263	72262	65946
365513	*	FU	E	*GT*	HE	65906	72265	72264	65947
365514	*	FU	E	*GT*	HE	65907	72267	72266	65948
365515	*	FU	E	*GT*	HE	65908	72269	72268	65949
365516	*	FU	E	*GT*	HE	65909	72271	72270	65950
365517		TL	E	*GT*	HE	65910	72273	72272	65951
365518		FU	E	*GT*	HE	65911	72275	72274	65952
365519		TL	E	*GT*	HE	65912	72277	72276	65953
365520		TL	E	*GT*	HE	65913	72279	72278	65954
365521		FU	E	*GT*	HE	65914	72281	72280	65955
365522		TL	E	*GT*	HE	65915	72283	72282	65956
365523		TL	E	*GT*	HE	65916	72285	72284	65957
365524		FU	E	*GT*	HE	65917	72287	72286	65958
365525		TL	E	*GT*	HE	65918	72289	72288	65959

365527	FU	E	GT	HE	65920	72293	72292	65961
365528	TL	E	GT	HE	65921	72295	72294	65962
365529	FU	E	GT	HE	65922	72297	72296	65963
365530	FU	E	GT	HE	65923	72299	72298	65964
365531	AL	E	GT	HE	65924	72301	72300	65965
365532	FU	E	GT	HE	65925	72303	72302	65966
365533	TL	E	GT	HE	65926	72305	72304	65967
365534	FU	E	GT	HE	65927	72307	72306	65968
365535	FU	E	GT	HE	65928	72309	72308	65969
365536	FU	E	GT	HE	65929	72311	72310	65970
365537	FU	E	GT	HE	65930	72313	72312	65971
365538	FU	E	GT	HE	65931	72315	72314	65972
365539	FU	E	GT	HE	65932	72317	72316	65973
365540	TL	E	GT	HE	65933	72319	72318	65974
365541	FU	E	GT	HE	65934	72321	72320	65975

Names (carried on each DMCO):

365506 The Royston Express
365513 Hornsey Depot
365514 Captain George Vancouver
365517 Supporting Red Balloon
365518 The Fenman
365527 Robert Stripe Passengers' Champion
365530 The Intalink Partnership promoting integrated transport in
        Hertfordshire since 1999
365533 Max Appeal
365536 Rufus Barnes Chief Executive of London TravelWatch for 25 years
365537 Daniel Edwards (1974–2010) Cambridge Driver

CLASS 375 ELECTROSTAR
ADTRANZ/BOMBARDIER DERBY

Express and outer suburban units.

Formations: Various.
Systems: 25 kV AC overhead/750 V DC third rail (some third rail only with
provision for retro-fitting of AC equipment).
Construction: Welded aluminium alloy underframe, sides and roof with
steel ends. All sections bolted together.
Traction Motors: Two Adtranz asynchronous of 250 kW.
Wheel Arrangement: 2-Bo (+ 2-Bo) + 2-2 + Bo-2.
Braking: Disc & regenerative.   **Dimensions:** 20.40/19.99 x 2.80 m.
Bogies: Adtranz P3-25/T3-25.   **Couplers:** Dellner 12.
Gangways: Throughout.   **Control System:** IGBT Inverter.
Doors: Sliding plug.   **Maximum Speed:** 100 mph.
Heating & ventilation: Air conditioning.
Seating Layout: 1: 2+2 facing/unidirectional (seats behind drivers cab in each
DMCO). 2: 2+2 facing/unidirectional (except 375/9 – 3+2 facing/unidirectional).
Multiple Working: Within class and with Classes 376, 377, 378 and 379.

Class 375/3. Express units. 750 V DC only. DMCO–TSO–DMCO.

DMCO(A). Bombardier Derby 2001–02. 12/48. 43.8 t.
TSO. Bombardier Derby 2001–02. –/56 1TD 2W. 35.5 t.
DMCO(B). Bombardier Derby 2001–02. 12/48. 43.8 t.

375301	**CN**	E	*SE*	RM	67921	74351	67931
375302	**CN**	E	*SE*	RM	67922	74352	67932
375303	**CN**	E	*SE*	RM	67923	74353	67933
375304	**CN**	E	*SE*	RM	67924	74354	67934
375305	**CN**	E	*SE*	RM	67925	74355	67935
375306	**CN**	E	*SE*	RM	67926	74356	67936
375307	**CN**	E	*SE*	RM	67927	74357	67937
375308	**SE**	E	*SE*	RM	67928	74358	67938
375309	**CN**	E	*SE*	RM	67929	74359	67939
375310	**CN**	E	*SE*	RM	67930	74360	67940

Name (carried on TSO): 375304 Medway Valley Line 1856–2006

Class 375/6. Express units. 25 kV AC/750 V DC. DMCO–MSO–PTSO–DMCO.

DMCO(A). Adtranz Derby 1999–2001. 12/48. 46.2 t.
MSO. Adtranz Derby 1999–2001. –/66 1T. 40.5 t.
PTSO. Adtranz Derby 1999–2001. –/56 1TD 2W. 40.7 t.
DMCO(B). Adtranz Derby 1999–2001. 12/48. 46.2 t.

375601	**CN**	E	*SE*	RM	67801	74251	74201	67851
375602	**CN**	E	*SE*	RM	67802	74252	74202	67852
375603	**CN**	E	*SE*	RM	67803	74253	74203	67853
375604	**CN**	E	*SE*	RM	67804	74254	74204	67854
375605	**CN**	E	*SE*	RM	67805	74255	74205	67855
375606	**CN**	E	*SE*	RM	67806	74256	74206	67856
375607	**CN**	E	*SE*	RM	67807	74257	74207	67857
375608	**CN**	E	*SE*	RM	67808	74258	74208	67858
375609	**SE**	E	*SE*	RM	67809	74259	74209	67859
375610	**CN**	E	*SE*	RM	67810	74260	74210	67860
375611	**CN**	E	*SE*	RM	67811	74261	74211	67861
375612	**CN**	E	*SE*	RM	67812	74262	74212	67862
375613	**CN**	E	*SE*	RM	67813	74263	74213	67863
375614	**CN**	E	*SE*	RM	67814	74264	74214	67864
375615	**CN**	E	*SE*	RM	67815	74265	74215	67865
375616	**CN**	E	*SE*	RM	67816	74266	74216	67866
375617	**CN**	E	*SE*	RM	67817	74267	74217	67867
375618	**CN**	E	*SE*	RM	67818	74268	74218	67868
375619	**CN**	E	*SE*	RM	67819	74269	74219	67869
375620	**CN**	E	*SE*	RM	67820	74270	74220	67870
375621	**CN**	E	*SE*	RM	67821	74271	74221	67871
375622	**CN**	E	*SE*	RM	67822	74272	74222	67872
375623	**CN**	E	*SE*	RM	67823	74273	74223	67873
375624	**SE**	E	*SE*	RM	67824	74274	74224	67874
375625	**CN**	E	*SE*	RM	67825	74275	74225	67875
375626	**CN**	E	*SE*	RM	67826	74276	74226	67876
375627	**CN**	E	*SE*	RM	67827	74277	74227	67877
375628	**CN**	E	*SE*	RM	67828	74278	74228	67878

| 375629 | **CN** | E | *SE* | RM | 67829 | 74279 | 74229 | 67879 |
| 375630 | **CN** | E | *SE* | RM | 67830 | 74280 | 74230 | 67880 |

Names (carried on one side of each MSO or PTSO):

375608 Bromley Travelwise
375610 Royal Tunbridge Wells
375611 Dr. William Harvey

375619 Driver John Neve
375623 Hospice in the Weald

Class 375/7. Express units. 750 V DC only. DMCO–MSO–TSO–DMCO.

DMCO(A). Bombardier Derby 2001–02. 12/48. 43.8 t.
MSO. Bombardier Derby 2001–02. –/66 1T. 36.4 t.
TSO. Bombardier Derby 2001–02. –/56 1TD 2W. 34.1 t.
DMCO(B). Bombardier Derby 2001–02. 12/48. 43.8 t.

375701	**CN**	E	*SE*	RM	67831	74281	74231	67881
375702	**CN**	E	*SE*	RM	67832	74282	74232	67882
375703	**CN**	E	*SE*	RM	67833	74283	74233	67883
375704	**CN**	E	*SE*	RM	67834	74284	74234	67884
375705	**SE**	E	*SE*	RM	67835	74285	74235	67885
375706	**CN**	E	*SE*	RM	67836	74286	74236	67886
375707	**CN**	E	*SE*	RM	67837	74287	74237	67887
375708	**CN**	E	*SE*	RM	67838	74288	74238	67888
375709	**CN**	E	*SE*	RM	67839	74289	74239	67889
375710	**CN**	E	*SE*	RM	67840	74290	74240	67890
375711	**CN**	E	*SE*	RM	67841	74291	74241	67891
375712	**CN**	E	*SE*	RM	67842	74292	74242	67892
375713	**CN**	E	*SE*	RM	67843	74293	74243	67893
375714	**CN**	E	*SE*	RM	67844	74294	74244	67894
375715	**CN**	E	*SE*	RM	67845	74295	74245	67895

Name (carried on one side of each MSO or TSO):

375701 Kent Air Ambulance Explorer

Class 375/8. Express units. 750 V DC only. DMCO–MSO–TSO–DMCO.

375 801–820 are fitted with de-icing equipment. TSO weighs 36.5 t.

DMCO(A). Bombardier Derby 2004. 12/48. 43.3 t.
MSO. Bombardier Derby 2004. –/66 1T. 39.8 t.
TSO. Bombardier Derby 2004. –/52 1TD 2W. 35.9 t.
DMCO(B). Bombardier Derby 2004. 12/52. 43.3 t.

375801	**CN**	E	*SE*	RM	73301	79001	78201	73701
375802	**CN**	E	*SE*	RM	73302	79002	78202	73702
375803	**CN**	E	*SE*	RM	73303	79003	78203	73703
375804	**CN**	E	*SE*	RM	73304	79004	78204	73704
375805	**CN**	E	*SE*	RM	73305	79005	78205	73705
375806	**CN**	E	*SE*	RM	73306	79006	78206	73706
375807	**CN**	E	*SE*	RM	73307	79007	78207	73707
375808	**CN**	E	*SE*	RM	73308	79008	78208	73708
375809	**CN**	E	*SE*	RM	73309	79009	78209	73709
375810	**CN**	E	*SE*	RM	73310	79010	78210	73710
375811	**CN**	E	*SE*	RM	73311	79011	78211	73711

375 812	**CN**	E	*SE*	RM	73312	79012	78212	73712
375 813	**CN**	E	*SE*	RM	73313	79013	78213	73713
375 814	**CN**	E	*SE*	RM	73314	79014	78214	73714
375 815	**CN**	E	*SE*	RM	73315	79015	78215	73715
375 816	**CN**	E	*SE*	RM	73316	79016	78216	73716
375 817	**CN**	E	*SE*	RM	73317	79017	78217	73717
375 818	**CN**	E	*SE*	RM	73318	79018	78218	73718
375 819	**CN**	E	*SE*	RM	73319	79019	78219	73719
375 820	**CN**	E	*SE*	RM	73320	79020	78220	73720
375 821	**CN**	E	*SE*	RM	73321	79021	78221	73721
375 822	**CN**	E	*SE*	RM	73322	79022	78222	73722
375 823	**CN**	E	*SE*	RM	73323	79023	78223	73723
375 824	**CN**	E	*SE*	RM	73324	79024	78224	73724
375 825	**CN**	E	*SE*	RM	73325	79025	78225	73725
375 826	**CN**	E	*SE*	RM	73326	79026	78226	73726
375 827	**CN**	E	*SE*	RM	73327	79027	78227	73727
375 828	**CN**	E	*SE*	RM	73328	79028	78228	73728
375 829	**CN**	E	*SE*	RM	73329	79029	78229	73729
375 830	**CN**	E	*SE*	RM	73330	79030	78230	73730

Name (carried on one side of each MSO or TSO):

375 830 City of London

Class 375/9. Outer suburban units. 750 V DC only. DMCO–MSO–TSO–DMCO.

DMCO(A). Bombardier Derby 2003–04. 12/59. 43.4 t.
MSO. Bombardier Derby 2003–04. –/73 1T. 39.3 t.
TSO. Bombardier Derby 2003–04. –/59 1TD 2W. 35.6 t.
DMCO(B). Bombardier Derby 2003–04. 12/59. 43.4 t.

375 901	**CN**	E	*SE*	RM	73331	79031	79061	73731
375 902	**CN**	E	*SE*	RM	73332	79032	79062	73732
375 903	**CN**	E	*SE*	RM	73333	79033	79063	73733
375 904	**CN**	E	*SE*	RM	73334	79034	79064	73734
375 905	**CN**	E	*SE*	RM	73335	79035	79065	73735
375 906	**CN**	E	*SE*	RM	73336	79036	79066	73736
375 907	**CN**	E	*SE*	RM	73337	79037	79067	73737
375 908	**CN**	E	*SE*	RM	73338	79038	79068	73738
375 909	**CN**	E	*SE*	RM	73339	79039	79069	73739
375 910	**CN**	E	*SE*	RM	73340	79040	79070	73740
375 911	**CN**	E	*SE*	RM	73341	79041	79071	73741
375 912	**CN**	E	*SE*	RM	73342	79042	79072	73742
375 913	**CN**	E	*SE*	RM	73343	79043	79073	73743
375 914	**CN**	E	*SE*	RM	73344	79044	79074	73744
375 915	**CN**	E	*SE*	RM	73345	79045	79075	73745
375 916	**CN**	E	*SE*	RM	73346	79046	79076	73746
375 917	**CN**	E	*SE*	RM	73347	79047	79077	73747
375 918	**CN**	E	*SE*	RM	73348	79048	79078	73748
375 919	**CN**	E	*SE*	RM	73349	79049	79079	73749
375 920	**CN**	E	*SE*	RM	73350	79050	79080	73750
375 921	**CN**	E	*SE*	RM	73351	79051	79081	73751
375 922	**CN**	E	*SE*	RM	73352	79052	79082	73752

375923	CN	E	SE	RM	73353	79053	79083	73753
375924	CN	E	SE	RM	73354	79054	79084	73754
375925	CN	E	SE	RM	73355	79055	79085	73755
375926	CN	E	SE	RM	73356	79056	79086	73756
375927	CN	E	SE	RM	73357	79057	79087	73757

CLASS 376   ELECTROSTAR   BOMBARDIER DERBY

Inner suburban units.

Formation: DMSO–MSO–TSO–MSO–DMSO.
System: 750 V DC third rail.
Construction: Welded aluminium alloy underframe, sides and roof with steel ends. All sections bolted together.
Traction Motors: Two Bombardier asynchronous of 200 kW.
Wheel Arrangement: 2-Bo + 2-Bo + 2-2 + Bo-2 + Bo-2.
Braking: Disc & regenerative.   **Dimensions:** 20.40/19.99 x 2.80 m.
Bogies: Bombardier P3-25/T3-25.   **Couplers:** Dellner 12.
Gangways: Within unit.   **Control System:** IGBT Inverter.
Doors: Sliding.   **Maximum Speed:** 75 mph.
Heating & ventilation: Pressure heating and ventilation.
Seating Layout: 2+2 low density facing.
Multiple Working: Within class and with Classes 375, 377, 378 and 379.

DMSO(A). Bombardier Derby 2004–05. –/36(6) 1W. 42.1 t.
MSO. Bombardier Derby 2004–05. –/48. 36.2 t.
TSO. Bombardier Derby 2004–05. –/48. 36.3 t.
DMSO(B). Bombardier Derby 2004–05. –/36(6) 1W. 42.1 t.

376001	CN	E	SE	SG	61101	63301	64301	63501	61601
376002	CN	E	SE	SG	61102	63302	64302	63502	61602
376003	CN	E	SE	SG	61103	63303	64303	63503	61603
376004	CN	E	SE	SG	61104	63304	64304	63504	61604
376005	CN	E	SE	SG	61105	63305	64305	63505	61605
376006	CN	E	SE	SG	61106	63306	64306	63506	61606
376007	CN	E	SE	SG	61107	63307	64307	63507	61607
376008	CN	E	SE	SG	61108	63308	64308	63508	61608
376009	CN	E	SE	SG	61109	63309	64309	63509	61609
376010	CN	E	SE	SG	61110	63310	64310	63510	61610
376011	CN	E	SE	SG	61111	63311	64311	63511	61611
376012	CN	E	SE	SG	61112	63312	64312	63512	61612
376013	CN	E	SE	SG	61113	63313	64313	63513	61613
376014	CN	E	SE	SG	61114	63314	64314	63514	61614
376015	CN	E	SE	SG	61115	63315	64315	63515	61615
376016	CN	E	SE	SG	61116	63316	64316	63516	61616
376017	CN	E	SE	SG	61117	63317	64317	63517	61617
376018	CN	E	SE	SG	61118	63318	64318	63518	61618
376019	CN	E	SE	SG	61119	63319	64319	63519	61619
376020	CN	E	SE	SG	61120	63320	64320	63520	61620
376021	CN	E	SE	SG	61121	63321	64321	63521	61621
376022	CN	E	SE	SG	61122	63322	64322	63522	61622
376023	CN	E	SE	SG	61123	63323	64323	63523	61623

376024	**CN**	E	*SE*	SG	61124	63324	64324	63524	61624
376025	**CN**	E	*SE*	SG	61125	63325	64325	63525	61625
376026	**CN**	E	*SE*	SG	61126	63326	64326	63526	61626
376027	**CN**	E	*SE*	SG	61127	63327	64327	63527	61627
376028	**CN**	E	*SE*	SG	61128	63328	64328	63528	61628
376029	**CN**	E	*SE*	SG	61129	63329	64329	63529	61629
376030	**CN**	E	*SE*	SG	61130	63330	64330	63530	61630
376031	**CN**	E	*SE*	SG	61131	63331	64331	63531	61631
376032	**CN**	E	*SE*	SG	61132	63332	64332	63532	61632
376033	**CN**	E	*SE*	SG	61133	63333	64333	63533	61633
376034	**CN**	E	*SE*	SG	61134	63334	64334	63534	61634
376035	**CN**	E	*SE*	SG	61135	63335	64335	63535	61635
376036	**CN**	E	*SE*	SG	61136	63336	64336	63536	61636

CLASS 377   ELECTROSTAR   BOMBARDIER DERBY

Express and outer suburban units.

Formations: Various.
Systems: 25 kV AC overhead/750 V DC third rail or third rail only with provision for retro-fitting of AC equipment.
Construction: Welded aluminium alloy underframe, sides and roof with steel ends. All sections bolted together.
Traction Motors: Two Bombardier asynchronous of 250 kW.
Wheel Arrangement: 2-Bo(+ 2-Bo) + 2-2 + Bo-2.
Braking: Disc & regenerative.　　　**Dimensions:** 20.39/20.00 x 2.80 m.
Bogies: Bombardier P3-25/T3-25.　　**Couplers:** Dellner 12.
Gangways: Throughout.　　　　　　**Control System:** IGBT Inverter.
Doors: Sliding plug.　　　　　　　**Maximum Speed:** 100 mph.
Heating & ventilation: Air conditioning.
Seating Layout: Various, see sub-class headings.
Multiple Working: Within class and with Classes 375, 376, 378 and 379.

Class 377/1. 750 V DC only. DMCO–MSO–TSO–DMCO.
Seating layout: 1: 2+2 facing/unidirectional, 2: 2+2 facing/unidirectional (377 101–119), 3+2 (middle cars) and 2+2 (end cars) facing/unidirectional (377 120–164).

DMCO(A). Bombardier Derby 2002–03. 12/48 (s 12/56). 44.8 t.
MSO. Bombardier Derby 2002–03. –/62 (s –/70, t –/69). 1T. 39.0 t.
TSO. Bombardier Derby 2002–03. –/52 (s –/60, t –/57). 1TD 2W. 35.4 t.
DMCO(B). Bombardier Derby 2002–03. 12/48 (s 12/56). 43.4 t.

377101	**SN**	P	*SN*	BI	78501	77101	78901	78701
377102	**SN**	P	*SN*	BI	78502	77102	78902	78702
377103	**SN**	P	*SN*	BI	78503	77103	78903	78703
377104	**SN**	P	*SN*	BI	78504	77104	78904	78704
377105	**SN**	P	*SN*	BI	78505	77105	78905	78705
377106	**SN**	P	*SN*	BI	78506	77106	78906	78706
377107	**SN**	P	*SN*	BI	78507	77107	78907	78707
377108	**SN**	P	*SN*	BI	78508	77108	78908	78708
377109	**SN**	P	*SN*	BI	78509	77109	78909	78709
377110	**SN**	P	*SN*	BI	78510	77110	78910	78710

377 111		**SN**	P	*SN*	BI	78511	77111	78911	78711
377 112		**SN**	P	*SN*	BI	78512	77112	78912	78712
377 113		**SN**	P	*SN*	BI	78513	77113	78913	78713
377 114		**SN**	P	*SN*	BI	78514	77114	78914	78714
377 115		**SN**	P	*SN*	BI	78515	77115	78915	78715
377 116		**SN**	P	*SN*	BI	78516	77116	78916	78716
377 117		**SN**	P	*SN*	BI	78517	77117	78917	78717
377 118		**SN**	P	*SN*	BI	78518	77118	78918	78718
377 119		**SN**	P	*SN*	BI	78519	77119	78919	78719
377 120	s	**SN**	P	*SN*	BI	78520	77120	78920	78720
377 121	s	**SN**	P	*SN*	BI	78521	77121	78921	78721
377 122	s	**SN**	P	*SN*	BI	78522	77122	78922	78722
377 123	s	**SN**	P	*SN*	BI	78523	77123	78923	78723
377 124	s	**SN**	P	*SN*	BI	78524	77124	78924	78724
377 125	s	**SN**	P	*SN*	BI	78525	77125	78925	78725
377 126	s	**SN**	P	*SN*	BI	78526	77126	78926	78726
377 127	s	**SN**	P	*SN*	BI	78527	77127	78927	78727
377 128	s	**SN**	P	*SN*	BI	78528	77128	78928	78728
377 129	s	**SN**	P	*SN*	BI	78529	77129	78929	78729
377 130	s	**SN**	P	*SN*	BI	78530	77130	78930	78730
377 131	s	**SN**	P	*SN*	BI	78531	77131	78931	78731
377 132	s	**SN**	P	*SN*	BI	78532	77132	78932	78732
377 133	s	**SN**	P	*SN*	BI	78533	77133	78933	78733
377 134	s	**SN**	P	*SN*	BI	78534	77134	78934	78734
377 135	s	**SN**	P	*SN*	BI	78535	77135	78935	78735
377 136	s	**SN**	P	*SN*	BI	78536	77136	78936	78736
377 137	s	**SN**	P	*SN*	BI	78537	77137	78937	78737
377 138	s	**SN**	P	*SN*	BI	78538	77138	78938	78738
377 139	s	**SN**	P	*SN*	BI	78539	77139	78939	78739
377 140	t	**SN**	P	*SN*	BI	78540	77140	78940	78740
377 141	t	**SN**	P	*SN*	BI	78541	77141	78941	78741
377 142	t	**SN**	P	*SN*	BI	78542	77142	78942	78742
377 143	t	**SN**	P	*SN*	BI	78543	77143	78943	78743
377 144	t	**SN**	P	*SN*	BI	78544	77144	78944	78744
377 145	t	**SN**	P	*SN*	BI	78545	77145	78945	78745
377 146	t	**SN**	P	*SN*	BI	78546	77146	78946	78746
377 147	t	**SN**	P	*SN*	BI	78547	77147	78947	78747
377 148	t	**SN**	P	*SN*	BI	78548	77148	78948	78748
377 149	t	**SN**	P	*SN*	BI	78549	77149	78949	78749
377 150	t	**SN**	P	*SN*	BI	78550	77150	78950	78750
377 151	t	**SN**	P	*SN*	BI	78551	77151	78951	78751
377 152	t	**SN**	P	*SN*	BI	78552	77152	78952	78752
377 153	t	**SN**	P	*SN*	BI	78553	77153	78953	78753
377 154	t	**SN**	P	*SN*	BI	78554	77154	78954	78754
377 155	t	**SN**	P	*SN*	BI	78555	77155	78955	78755
377 156	t	**SN**	P	*SN*	BI	78556	77156	78956	78756
377 157	t	**SN**	P	*SN*	BI	78557	77157	78957	78757
377 158	t	**SN**	P	*SN*	BI	78558	77158	78958	78758
377 159	t	**SN**	P	*SN*	BI	78559	77159	78959	78759
377 160	t	**SN**	P	*SN*	BI	78560	77160	78960	78760
377 161	t	**SN**	P	*SN*	BI	78561	77161	78961	78761

377 162	t	**SN**	P	*SN*	BI	78562	77162	78962	78762
377 163	t	**SN**	P	*SN*	BI	78563	77163	78963	78763
377 164	t	**SN**	P	*SN*	BI	78564	77164	78964	78764

Class 377/2. 25 kV AC/750 V DC. DMCO–MSO–PTSO–DMCO. Dual-voltage units. 377 207–215 are sub-leased from Southern to Govia Thameslink Railway.
Seating layout: 1: 2+2 facing/unidirectional, 2: 2+2 and 3+2 facing/unidirectional (3+2 seating in middle cars only).

DMCO(A). Bombardier Derby 2003–04. 12/48. 44.2 t.
MSO. Bombardier Derby 2003–04. –/69 1T. 39.8 t.
PTSO. Bombardier Derby 2003–04. –/57 1TD 2W. 40.1 t.
DMCO(B). Bombardier Derby 2003–04. 12/48. 44.2 t.

377 201	**SN**	P	*SN*	SU	78571	77171	78971	78771
377 202	**SN**	P	*SN*	SU	78572	77172	78972	78772
377 203	**SN**	P	*SN*	SU	78573	77173	78973	78773
377 204	**SN**	P	*SN*	SU	78574	77174	78974	78774
377 205	**SN**	P	*SN*	SU	78575	77175	78975	78775
377 206	**SN**	P	*SN*	SU	78576	77176	78976	78776
377 207	**FU**	P	*GT*	BF	78577	77177	78977	78777
377 208	**SN**	P	*GT*	BF	78578	77178	78978	78778
377 209	**SN**	P	*GT*	BF	78579	77179	78979	78779
377 210	**SN**	P	*GT*	BF	78580	77180	78980	78780
377 211	**FU**	P	*GT*	BF	78581	77181	78981	78781
377 212	**FU**	P	*GT*	BF	78582	77182	78982	78782
377 213	**SN**	P	*GT*	BF	78583	77183	78983	78783
377 214	**SN**	P	*GT*	BF	78584	77184	78984	78784
377 215	**SN**	P	*GT*	BF	78585	77185	78985	78785

Class 377/3. 750 V DC only. DMCO–TSO–DMCO.
Seating Layout: 1: 2+2 facing/unidirectional, 2: 2+2 facing/unidirectional.

Units built as Class 375, but renumbered in the Class 377/3 range when fitted with Dellner couplers.

DMCO(A). Bombardier Derby 2001–02. 12/48. 43.5 t.
TSO. Bombardier Derby 2001–02. –/56 1TD 2W. 35.4 t.
DMCO(B). Bombardier Derby 2001–02. 12/48. 43.5 t.

377 301	(375 311)	**SN**	P	*SN*	SU	68201	74801	68401
377 302	(375 312)	**SN**	P	*SN*	SU	68202	74802	68402
377 303	(375 313)	**SN**	P	*SN*	SU	68203	74803	68403
377 304	(375 314)	**SN**	P	*SN*	SU	68204	74804	68404
377 305	(375 315)	**SN**	P	*SN*	SU	68205	74805	68405
377 306	(375 316)	**SN**	P	*SN*	SU	68206	74806	68406
377 307	(375 317)	**SN**	P	*SN*	SU	68207	74807	68407
377 308	(375 318)	**SN**	P	*SN*	SU	68208	74808	68408
377 309	(375 319)	**SN**	P	*SN*	SU	68209	74809	68409
377 310	(375 320)	**SN**	P	*SN*	SU	68210	74810	68410
377 311	(375 321)	**SN**	P	*SN*	SU	68211	74811	68411
377 312	(375 322)	**SN**	P	*SN*	SU	68212	74812	68412
377 313	(375 323)	**SN**	P	*SN*	SU	68213	74813	68413

377 314	(375 324)	**SN**	P	*SN*	SU	68214	74814	68414
377 315	(375 325)	**SN**	P	*SN*	SU	68215	74815	68415
377 316	(375 326)	**SN**	P	*SN*	SU	68216	74816	68416
377 317	(375 327)	**SN**	P	*SN*	SU	68217	74817	68417
377 318	(375 328)	**SN**	P	*SN*	SU	68218	74818	68418
377 319	(375 329)	**SN**	P	*SN*	SU	68219	74819	68419
377 320	(375 330)	**SN**	P	*SN*	SU	68220	74820	68420
377 321	(375 331)	**SN**	P	*SN*	SU	68221	74821	68421
377 322	(375 332)	**SN**	P	*SN*	SU	68222	74822	68422
377 323	(375 333)	**SN**	P	*SN*	SU	68223	74823	68423
377 324	(375 334)	**SN**	P	*SN*	SU	68224	74824	68424
377 325	(375 335)	**SN**	P	*SN*	SU	68225	74825	68425
377 326	(375 336)	**SN**	P	*SN*	SU	68226	74826	68426
377 327	(375 337)	**SN**	P	*SN*	SU	68227	74827	68427
377 328	(375 338)	**SN**	P	*SN*	SU	68228	74828	68428

Class 377/4. 750 V DC only. DMCO–MSO–TSO–DMCO.
Seating Layout: 1: 2+2 facing/two seats longitudinal, 2: 2+2 and 3+2 facing/unidirectional (3+2 seating in middle cars only).

DMCO(A). Bombardier Derby 2004–05. 10/48. 43.1 t.
MSO. Bombardier Derby 2004–05. –/69 1T. 39.3 t.
TSO. Bombardier Derby 2004–05. –/56 1TD 2W. 35.3 t.
DMCO(B). Bombardier Derby 2004–05. 10/48. 43.2 t.

377 401	**SN**	P	*SN*	BI	73401	78801	78601	73801
377 402	**SN**	P	*SN*	BI	73402	78802	78602	73802
377 403	**SN**	P	*SN*	BI	73403	78803	78603	73803
377 404	**SN**	P	*SN*	BI	73404	78804	78604	73804
377 405	**SN**	P	*SN*	BI	73405	78805	78605	73805
377 406	**SN**	P	*SN*	BI	73406	78806	78606	73806
377 407	**SN**	P	*SN*	BI	73407	78807	78607	73807
377 408	**SN**	P	*SN*	BI	73408	78808	78608	73808
377 409	**SN**	P	*SN*	BI	73409	78809	78609	73809
377 410	**SN**	P	*SN*	BI	73410	78810	78610	73810
377 411	**SN**	P	*SN*	BI	73411	78811	78611	73811
377 412	**SN**	P	*SN*	BI	73412	78812	78612	73812
377 413	**SN**	P	*SN*	BI	73413	78813	78613	73813
377 414	**SN**	P	*SN*	BI	73414	78814	78614	73814
377 415	**SN**	P	*SN*	BI	73415	78815	78615	73815
377 416	**SN**	P	*SN*	BI	73416	78816	78616	73816
377 417	**SN**	P	*SN*	BI	73417	78817	78617	73817
377 418	**SN**	P	*SN*	BI	73418	78818	78618	73818
377 419	**SN**	P	*SN*	BI	73419	78819	78619	73819
377 420	**SN**	P	*SN*	BI	73420	78820	78620	73820
377 421	**SN**	P	*SN*	BI	73421	78821	78621	73821
377 422	**SN**	P	*SN*	BI	73422	78822	78622	73822
377 423	**SN**	P	*SN*	BI	73423	78823	78623	73823
377 424	**SN**	P	*SN*	BI	73424	78824	78624	73824
377 425	**SN**	P	*SN*	BI	73425	78825	78625	73825
377 426	**SN**	P	*SN*	BI	73426	78826	78626	73826
377 427	**SN**	P	*SN*	BI	73427	78827	78627	73827

377 428	**SN**	P	*SN*	Bl	73428	78828	78628	73828
377 429	**SN**	P	*SN*	Bl	73429	78829	78629	73829
377 430	**SN**	P	*SN*	Bl	73430	78830	78630	73830
377 431	**SN**	P	*SN*	Bl	73431	78831	78631	73831
377 432	**SN**	P	*SN*	Bl	73432	78832	78632	73832
377 433	**SN**	P	*SN*	Bl	73433	78833	78633	73833
377 434	**SN**	P	*SN*	Bl	73434	78834	78634	73834
377 435	**SN**	P	*SN*	Bl	73435	78835	78635	73835
377 436	**SN**	P	*SN*	Bl	73436	78836	78636	73836
377 437	**SN**	P	*SN*	Bl	73437	78837	78637	73837
377 438	**SN**	P	*SN*	Bl	73438	78838	78638	73838
377 439	**SN**	P	*SN*	Bl	73439	78839	78639	73839
377 440	**SN**	P	*SN*	Bl	73440	78840	78640	73840
377 441	**SN**	P	*SN*	Bl	73441	78841	78641	73841
377 442	**SN**	P	*SN*	Bl	73442	78842	78642	73842
377 443	**SN**	P	*SN*	Bl	73443	78843	78643	73843
377 444	**SN**	P	*SN*	Bl	73444	78844	78644	73844
377 445	**SN**	P	*SN*	Bl	73445	78845	78645	73845
377 446	**SN**	P	*SN*	Bl	73446	78846	78646	73846
377 447	**SN**	P	*SN*	Bl	73447	78847	78647	73847
377 448	**SN**	P	*SN*	Bl	73448	78848	78648	73848
377 449	**SN**	P	*SN*	Bl	73449	78849	78649	73849
377 450	**SN**	P	*SN*	Bl	73450	78850	78650	73850
377 451	**SN**	P	*SN*	Bl	73451	78851	78651	73851
377 452	**SN**	P	*SN*	Bl	73452	78852	78652	73852
377 453	**SN**	P	*SN*	Bl	73453	78853	78653	73853
377 454	**SN**	P	*SN*	Bl	73454	78854	78654	73854
377 455	**SN**	P	*SN*	Bl	73455	78855	78655	73855
377 456	**SN**	P	*SN*	Bl	73456	78856	78656	73856
377 457	**SN**	P	*SN*	Bl	73457	78857	78657	73857
377 458	**SN**	P	*SN*	Bl	73458	78858	78658	73858
377 459	**SN**	P	*SN*	Bl	73459	78859	78659	73859
377 460	**SN**	P	*SN*	Bl	73460	78860	78660	73860
377 461	**SN**	P	*SN*	Bl	73461	78861	78661	73861
377 462	**SN**	P	*SN*	Bl	73462	78862	78662	73862
377 463	**SN**	P	*SN*	Bl	73463	78863	78663	73863
377 464	**SN**	P	*SN*	Bl	73464	78864	78664	73864
377 465	**SN**	P	*SN*	Bl	73465	78865	78665	73865
377 466	**SN**	P	*SN*	Bl	73466	78866	78666	73866
377 467	**SN**	P	*SN*	Bl	73467	78867	78667	73867
377 468	**SN**	P	*SN*	Bl	73468	78868	78668	73868
377 469	**SN**	P	*SN*	Bl	73469	78869	78669	73869
377 470	**SN**	P	*SN*	Bl	73470	78870	78670	73870
377 471	**SN**	P	*SN*	Bl	73471	78871	78671	73871
377 472	**SN**	P	*SN*	Bl	73472	78872	78672	73872
377 473	**SN**	P	*SN*	Bl	73473	78873	78673	73873
377 474	**SN**	P	*SN*	Bl	73474	78874	78674	73874
377 475	**SN**	P	*SN*	Bl	73475	78875	78675	73875

Class 377/5. 25 kV AC/750 V DC. DMCO–MSO–PTSO–DMCO. Dual voltage FCC units (sub-leased from Southern). Details as Class 377/2 unless stated.

DMCO(A). Bombardier Derby 2008–09. 10/48. 43.1 t.
MSO. Bombardier Derby 2008–09. –/69 1T. 40.3 t.
PTSO. Bombardier Derby 2008–09. –/56 1TD 2W. 40.6 t.
DMCO(B). Bombardier Derby 2008–09. 10/48. 43.1 t.

377 501	**FU**	P	*GT*	BF	73501	75901	74901	73601
377 502	**FU**	P	*GT*	BF	73502	75902	74902	73602
377 503	**FU**	P	*GT*	BF	73503	75903	74903	73603
377 504	**FU**	P	*GT*	BF	73504	75904	74904	73604
377 505	**FU**	P	*GT*	BF	73505	75905	74905	73605
377 506	**FU**	P	*GT*	BF	73506	75906	74906	73606
377 507	**FU**	P	*GT*	BF	73507	75907	74907	73607
377 508	**FU**	P	*GT*	BF	73508	75908	74908	73608
377 509	**FU**	P	*GT*	BF	73509	75909	74909	73609
377 510	**FU**	P	*GT*	BF	73510	75910	74910	73610
377 511	**FU**	P	*GT*	BF	73511	75911	74911	73611
377 512	**FU**	P	*GT*	BF	73512	75912	74912	73612
377 513	**FU**	P	*GT*	BF	73513	75913	74913	73613
377 514	**FU**	P	*GT*	BF	73514	75914	74914	73614
377 515	**FU**	P	*GT*	BF	73515	75915	74915	73615
377 516	**FU**	P	*GT*	BF	73516	75916	74916	73616
377 517	**FU**	P	*GT*	BF	73517	75917	74917	73617
377 518	**FU**	P	*GT*	BF	73518	75918	74918	73618
377 519	**FU**	P	*GT*	BF	73519	75919	74919	73619
377 520	**FU**	P	*GT*	BF	73520	75920	74920	73620
377 521	**FU**	P	*GT*	BF	73521	75921	74921	73621
377 522	**FU**	P	*GT*	BF	73522	75922	74922	73622
377 523	**FU**	P	*GT*	BF	73523	75923	74923	73623

Class 377/6. 750 V DC. DMSO–MSO–TSO–MSO–DMSO. Southern 5-car suburban units fitted with Fainsa seating. Technically the same as the 377/5s but using the slightly modified Class 379-style bodyshell.
Seating Layout: 2+2 facing/unidirectional.

DMSO. Bombardier Derby 2012–13. –/60. 44.7 t.
MSO. Bombardier Derby 2012–13. –/64 1T. 38.8 t.
TSO. Bombardier Derby 2012–13. –/46(2) 1TD 2W. 37.8 t.
MSO. Bombardier Derby 2012–13. –/66. 38.3 t.
DMSO. Bombardier Derby 2012–13. –/62. 44.7 t.

377 601	**SN**	P	*SN*	SU	70101	70201	70301	70401	70501
377 602	**SN**	P	*SN*	SU	70102	70202	70302	70402	70502
377 603	**SN**	P	*SN*	SU	70103	70203	70303	70403	70503
377 604	**SN**	P	*SN*	SU	70104	70204	70304	70404	70504
377 605	**SN**	P	*SN*	SU	70105	70205	70305	70405	70505
377 606	**SN**	P	*SN*	SU	70106	70206	70306	70406	70506
377 607	**SN**	P	*SN*	SU	70107	70207	70307	70407	70507
377 608	**SN**	P	*SN*	SU	70108	70208	70308	70408	70508
377 609	**SN**	P	*SN*	SU	70109	70209	70309	70409	70509
377 610	**SN**	P	*SN*	SU	70110	70210	70310	70410	70510
377 611	**SN**	P	*SN*	SU	70111	70211	70311	70411	70511

377612	**SN**	P	*SN*	SU	70112 70212 70312 70412 70512
377613	**SN**	P	*SN*	SU	70113 70213 70313 70413 70513
377614	**SN**	P	*SN*	SU	70114 70214 70314 70414 70514
377615	**SN**	P	*SN*	SU	70115 70215 70315 70415 70515
377616	**SN**	P	*SN*	SU	70116 70216 70316 70416 70516
377617	**SN**	P	*SN*	SU	70117 70217 70317 70417 70517
377618	**SN**	P	*SN*	SU	70118 70218 70318 70418 70518
377619	**SN**	P	*SN*	SU	70119 70219 70319 70419 70519
377620	**SN**	P	*SN*	SU	70120 70220 70320 70420 70520
377621	**SN**	P	*SN*	SU	70121 70221 70321 70421 70521
377622	**SN**	P	*SN*	SU	70122 70222 70322 70422 70522
377623	**SN**	P	*SN*	SU	70123 70223 70323 70423 70523
377624	**SN**	P	*SN*	SU	70124 70224 70324 70424 70524
377625	**SN**	P	*SN*	SU	70125 70225 70325 70425 70525
377626	**SN**	P	*SN*	SU	70126 70226 70326 70426 70526

Class 377/7. 25kV AC/750V DC. DMSO–MSO–TSO–MSO–DMSO. Dual voltage Southern units, used on both the South Croydon–Milton Keynes cross-London services and on suburban services alongside the 377/6s.

DMSO. Bombardier Derby 2013–14. –/60. 45.6 t.
MSO. Bombardier Derby 2013–14. –/64 1T. 41.0 t.
TSO. Bombardier Derby 2013–14. –/46(2) 1TD 2W. 40.9 t.
MSO. Bombardier Derby 2013–14. –/66. 39.6 t.
DMSO. Bombardier Derby 2013–14. –/62. 45.2 t.

377701	**SN**	P	*SN*	SU	65201 70601 65601 70701 65401
377702	**SN**	P	*SN*	SU	65202 70602 65602 70702 65402
377703	**SN**	P	*SN*	SU	65203 70603 65603 70703 65403
377704	**SN**	P	*SN*	SU	65204 70604 65604 70704 65404
377705	**SN**	P	*SN*	SU	65205 70605 65605 70705 65405
377706	**SN**	P	*SN*	SU	65206 70606 65606 70706 65406
377707	**SN**	P	*SN*	SU	65207 70607 65607 70707 65407
377708	**SN**	P	*SN*	SU	65208 70608 65608 70708 65408

CLASS 378   CAPITALSTAR   BOMBARDIER DERBY

57 Class 378 suburban Electrostars (designated Capitalstars by TfL).

Formation: DMSO–MSO–TSO–DMSO or DMSO–MSO–PTSO–DMSO.
System: Class 378/1 750 V DC third rail only. Class 378/2 25 kV AC overhead and 750 V DC third rail.
Construction: Welded aluminium alloy underframe, sides and roof with steel ends. All sections bolted together.
Traction Motors: Two Bombardier asynchronous of 200 kW.
Wheel Arrangement: 1A-Bo + 1A-Bo + 2-2 + Bo-1A.
Braking: Disc & regenerative.   **Dimensions:** 20.46/20.14 x 2.80 m.
Bogies: Bombardier P3-25/T3-25.   **Couplers:** Dellner 12.
Gangways: Within unit + end doors. **Control System:** IGBT Inverter.
Doors: Sliding.   **Maximum Speed:** 75 mph.
Heating & ventilation: Air conditioning.
Seating Layout: Longitudinal ("tube style") low density.
Multiple Working: Within class and with Classes 375, 376, 377 and 379.

57 extra vehicles are on order to make all units up to 5-cars. This will take place in 2014–15, starting with 378201 and then units 378135–154. The extra vehicles are numbered in the 38401–457 series and will be inserted between the TSO and DMSO(B).

Class 378/1. 750 V DC. DMSO–MSO–TSO–DMSO. Third rail only units used on East London Line services. Provision for retro-fitting as dual voltage.

378150–154 are fitted with de-icing equipment.

DMSO(A). Bombardier Derby 2009–10. –/36. 43.1 t.
MSO. Bombardier Derby 2009–10. –/40. 39.3 t.
TSO. Bombardier Derby 2009–10. –/34(6) 2W. 34.3 t.
DMSO(B). Bombardier Derby 2009–10. –/36. 42.7 t.

378135	**LO**	QW *LO*	NG	38035	38235	38335	38135
378136	**LO**	QW *LO*	NG	38036	38236	38336	38136
378137	**LO**	QW *LO*	NG	38037	38237	38337	38137
378138	**LO**	QW *LO*	NG	38038	38238	38338	38138
378139	**LO**	QW *LO*	NG	38039	38239	38339	38139
378140	**LO**	QW *LO*	NG	38040	38240	38340	38140
378141	**LO**	QW *LO*	NG	38041	38241	38341	38141
378142	**LO**	QW *LO*	NG	38042	38242	38342	38142
378143	**LO**	QW *LO*	NG	38043	38243	38343	38143
378144	**LO**	QW *LO*	NG	38044	38244	38344	38144
378145	**LO**	QW *LO*	NG	38045	38245	38345	38145
378146	**LO**	QW *LO*	NG	38046	38246	38346	38146
378147	**LO**	QW *LO*	NG	38047	38247	38347	38147
378148	**LO**	QW *LO*	NG	38048	38248	38348	38148
378149	**LO**	QW *LO*	NG	38049	38249	38349	38149
378150	**LO**	QW *LO*	NG	38050	38250	38350	38150
378151	**LO**	QW *LO*	NG	38051	38251	38351	38151
378152	**LO**	QW *LO*	NG	38052	38252	38352	38152
378153	**LO**	QW *LO*	NG	38053	38253	38353	38153
378154	**LO**	QW *LO*	NG	38054	38254	38354	38154

Class 378/2. 25 kV AC/750 V DC. DMSO–MSO–PTSO–(MSO)–DMSO. Dual voltage units mainly used on North London Railway services. 378201–224 built as 378001–024 (3-car units) and extended to 4-car units in 2010.

378216–220 are fitted with de-icing equipment.

Advertising livery: 378 211 and 378 221 Lycamobile (white).

DMSO(A). Bombardier Derby 2008–11. –/36. 43.4 t.
MSO. Bombardier Derby 2008–11. –/40. 39.6 t.
PTSO. Bombardier Derby 2008–11. –/34(6) 2W. 39.2 t.
DMSO(B). Bombardier Derby 2008–11. –/36. 43.1 t.

378201	**LO**	QW *LO*	NG	38001	38201	38301	38401	38101
378202	**LO**	QW *LO*	NG	38002	38202	38302		38102
378203	**LO**	QW *LO*	NG	38003	38203	38303		38103
378204	**LO**	QW *LO*	NG	38004	38204	38304		38104
378205	**LO**	QW *LO*	NG	38005	38205	38305		38105
378206	**LO**	QW *LO*	NG	38006	38206	38306		38106
378207	**LO**	QW *LO*	NG	38007	38207	38307		38107

378 208	**LO**	QW	*LO*	NG	38008 38208 38308		38108
378 209	**LO**	QW	*LO*	NG	38009 38209 38309		38109
378 210	**LO**	QW	*LO*	NG	38010 38210 38310		38110
378 211	**AL**	QW	*LO*	NG	38011 38211 38311		38111
378 212	**LO**	QW	*LO*	NG	38012 38212 38312		38112
378 213	**LO**	QW	*LO*	NG	38013 38213 38313		38113
378 214	**LO**	QW	*LO*	NG	38014 38214 38314		38114
378 215	**LO**	QW	*LO*	NG	38015 38215 38315		38115
378 216	**LO**	QW	*LO*	NG	38016 38216 38316		38116
378 217	**LO**	QW	*LO*	NG	38017 38217 38317		38117
378 218	**LO**	QW	*LO*	NG	38018 38218 38318		38118
378 219	**LO**	QW	*LO*	NG	38019 38219 38319		38119
378 220	**LO**	QW	*LO*	NG	38020 38220 38320		38120
378 221	**AL**	QW	*LO*	NG	38021 38221 38321		38121
378 222	**LO**	QW	*LO*	NG	38022 38222 38322		38122
378 223	**LO**	QW	*LO*	NG	38023 38223 38323		38123
378 224	**LO**	QW	*LO*	NG	38024 38224 38324		38124
378 225	**LO**	QW	*LO*	NG	38025 38225 38325		38125
378 226	**LO**	QW	*LO*	NG	38026 38226 38326		38126
378 227	**LO**	QW	*LO*	NG	38027 38227 38327		38127
378 228	**LO**	QW	*LO*	NG	38028 38228 38328		38128
378 229	**LO**	QW	*LO*	NG	38029 38229 38329		38129
378 230	**LO**	QW	*LO*	NG	38030 38230 38330		38130
378 231	**LO**	QW	*LO*	NG	38031 38231 38331		38131
378 232	**LO**	QW	*LO*	NG	38032 38232 38332		38132
378 233	**LO**	QW	*LO*	NG	38033 38233 38333		38133
378 234	**LO**	QW	*LO*	NG	38034 38234 38334		38134
378 255	**LO**	QW	*LO*	NG	38055 38255 38355		38155
378 256	**LO**	QW	*LO*	NG	38056 38256 38356		38156
378 257	**LO**	QW	*LO*	NG	38057 38257 38357		38157

Name (carried on 38033): 378 233   Ian Brown CBE

CLASS 379   ELECTROSTAR   BOMBARDIER DERBY

Express Electrostars used on Liverpool Street–Stansted Airport and Liverpool Street–Cambridge services.

Formation: DMSO–MSO–PTSO–DMCO.
System: 25 kV AC overhead.
Construction: Welded aluminium alloy underframe, sides and roof with steel ends. All sections bolted together.
Traction Motors: Two Bombardier asynchronous of 200 kW.
Wheel Arrangement: 2-Bo + 2-Bo + 2-2 + Bo-2.

Braking: Disc & regenerative.	**Dimensions:** 20.00 x 2.80 m.
Bogies: Bombardier P3-25/T3-25.	**Couplers:** Dellner 12.
Gangways: Throughout.	**Control System:** IGBT Inverter.
Doors: Sliding plug.	**Maximum Speed:** 100 mph.

Heating & ventilation: Air conditioning.
Seating Layout: 1: 2+1 facing. 2: 2+2 facing/unidirectional.
Multiple Working: Within class and with Classes 375, 376, 377 and 378.

379013 is being converted into a trial battery-electric hybrid unit, with the MSO vehicle being converted to a Trailer Battery Open.

DMSO. Bombardier Derby 2010–11. –/60. 42.1 t.
MSO. Bombardier Derby 2010–11. –/62 1T. 38.6 t.
PTSO. Bombardier Derby 2010–11. –/43(2) 1TD 2W. 40.9 t.
DMCO. Bombardier Derby 2010–11. 20/24. 42.3 t.

379001	**NC**	MQ	*GA*	IL	61201	61701	61901	62101
379002	**NC**	MQ	*GA*	IL	61202	61702	61902	62102
379003	**NC**	MQ	*GA*	IL	61203	61703	61903	62103
379004	**NC**	MQ	*GA*	IL	61204	61704	61904	62104
379005	**NC**	MQ	*GA*	IL	61205	61705	61905	62105
379006	**NC**	MQ	*GA*	IL	61206	61706	61906	62106
379007	**NC**	MQ	*GA*	IL	61207	61707	61907	62107
379008	**NC**	MQ	*GA*	IL	61208	61708	61908	62108
379009	**NC**	MQ	*GA*	IL	61209	61709	61909	62109
379010	**NC**	MQ	*GA*	IL	61210	61710	61910	62110
379011	**NC**	MQ	*GA*	IL	61211	61711	61911	62111
379012	**NC**	MQ	*GA*	IL	61212	61712	61912	62112
379013	**NC**	MQ	*GA*	IL	61213	61713	61913	62113
379014	**NC**	MQ	*GA*	IL	61214	61714	61914	62114
379015	**NC**	MQ	*GA*	IL	61215	61715	61915	62115
379016	**NC**	MQ	*GA*	IL	61216	61716	61916	62116
379017	**NC**	MQ	*GA*	IL	61217	61717	61917	62117
379018	**NC**	MQ	*GA*	IL	61218	61718	61918	62118
379019	**NC**	MQ	*GA*	IL	61219	61719	61919	62119
379020	**NC**	MQ	*GA*	IL	61220	61720	61920	62120
379021	**NC**	MQ	*GA*	IL	61221	61721	61921	62121
379022	**NC**	MQ	*GA*	IL	61222	61722	61922	62122
379023	**NC**	MQ	*GA*	IL	61223	61723	61923	62123
379024	**NC**	MQ	*GA*	IL	61224	61724	61924	62124
379025	**NC**	MQ	*GA*	IL	61225	61725	61925	62125
379026	**NC**	MQ	*GA*	IL	61226	61726	61926	62126
379027	**NC**	MQ	*GA*	IL	61227	61727	61927	62127
379028	**NC**	MQ	*GA*	IL	61228	61728	61928	62128
379029	**NC**	MQ	*GA*	IL	61229	61729	61929	62129
379030	**NC**	MQ	*GA*	IL	61230	61730	61930	62130

Names (carried on end cars):

379005	Stansted Express	379015	City of Cambridge
379011	Ely Cathedral	379025	Go Discover
379012	The West Anglian		

CLASS 380   DESIRO UK   SIEMENS

Used on Strathclyde area services and the Edinburgh–North Berwick route.

Formation: DMSO–PTSO–DMSO or DMSO–PTSO–TSO–DMSO.
System: 25 kV AC overhead.
Construction: Welded aluminium with steel ends.
Traction Motors: Four Siemens 1TB2016-0GB02 asynchronous of 250 kW.
Wheel Arrangement: Bo-Bo + 2-2 (+2-2) + Bo-Bo

Braking: Disc & regenerative.
Bogies: SGP SF5000.
Gangways: Throughout.
Doors: Sliding plug.
Heating & ventilation: Air conditioning.
Multiple Working: Within class.

Dimensions: 23.78/23.57 x 2.80 m.
Couplers: Voith.
Control System: IGBT Inverter.
Maximum Speed: 100 mph.
Seating Layout: 2+2 facing/unidirectional.

DMSO(A). Siemens Krefeld 2009–10. –/70. 45.0 t.
PTSO. Siemens Krefeld 2009–10. –/57(12) 1TD 2W. 42.7 t.
TSO. Siemens Krefeld 2009–10. –/74 1T. 34.8 t.
DMSO(B). Siemens Krefeld 2009–10. –/64(5). 44.9 t.

Class 380/0. 3-car units.

380 001	**SR**	E	*SR*	GW	38501	38601		38701
380 002	**SR**	E	*SR*	GW	38502	38602		38702
380 003	**SR**	E	*SR*	GW	38503	38603		38703
380 004	**SR**	E	*SR*	GW	38504	38604		38704
380 005	**SR**	E	*SR*	GW	38505	38605		38705
380 006	**SR**	E	*SR*	GW	38506	38606		38706
380 007	**SR**	E	*SR*	GW	38507	38607		38707
380 008	**SR**	E	*SR*	GW	38508	38608		38708
380 009	**SR**	E	*SR*	GW	38509	38609		38709
380 010	**SR**	E	*SR*	GW	38510	38610		38710
380 011	**SR**	E	*SR*	GW	38511	38611		38711
380 012	**SR**	E	*SR*	GW	38512	38612		38712
380 013	**SR**	E	*SR*	GW	38513	38613		38713
380 014	**SR**	E	*SR*	GW	38514	38614		38714
380 015	**SR**	E	*SR*	GW	38515	38615		38715
380 016	**SR**	E	*SR*	GW	38516	38616		38716
380 017	**SR**	E	*SR*	GW	38517	38617		38717
380 018	**SR**	E	*SR*	GW	38518	38618		38718
380 019	**SR**	E	*SR*	GW	38519	38619		38719
380 020	**SR**	E	*SR*	GW	38520	38620		38720
380 021	**SR**	E	*SR*	GW	38521	38621		38721
380 022	**SR**	E	*SR*	GW	38522	38622		38722

Class 380/1. 4-car units.

380 101	**SR**	E	*SR*	GW	38551	38651	38851	38751
380 102	**SR**	E	*SR*	GW	38552	38652	38852	38752
380 103	**SR**	E	*SR*	GW	38553	38653	38853	38753
380 104	**SR**	E	*SR*	GW	38554	38654	38854	38754
380 105	**SR**	E	*SR*	GW	38555	38655	38855	38755
380 106	**SR**	E	*SR*	GW	38556	38656	38856	38756
380 107	**SR**	E	*SR*	GW	38557	38657	38857	38757
380 108	**SR**	E	*SR*	GW	38558	38658	38858	38758
380 109	**SR**	E	*SR*	GW	38559	38659	38859	38759
380 110	**SR**	E	*SR*	GW	38560	38660	38860	38760
380 111	**SR**	E	*SR*	GW	38561	38661	38861	38761
380 112	**SR**	E	*SR*	GW	38562	38662	38862	38762
380 113	**SR**	E	*SR*	GW	38563	38663	38863	38763
380 114	**SR**	E	*SR*	GW	38564	38664	38864	38764
380 115	**SR**	E	*SR*	GW	38565	38665	38865	38765
380 116	**SR**	E	*SR*	GW	38566	38666	38866	38766

CLASS 387   ELECTROSTAR   BOMBARDIER DERBY

These 110 mph Electrostar units are on order to operate on the main Thameslink route until the new Class 700s are available. The Class 387s will be the first units to feature 6-digit vehicle numbers. Due to enter service from early 2015. Full details awaited.

An option exists for up to a further 35 4-car Class 387s.

Formation: DMSO–MSO–PTSO–DMSO.
System: 25 kV AC overhead and 750 V DC third rail.
Construction: Welded aluminium alloy underframe, sides and roof with steel ends. All sections bolted together.
Traction Motors:
Wheel Arrangement:

Braking: Disc & regenerative.	**Dimensions:**
Bogies:	**Couplers:**
Gangways: Throughout.	**Control System:** IGBT Inverter.
Doors: Sliding plug.	**Maximum Speed:** 110 mph.

Heating & ventilation: Air conditioning.
Seating Layout: 2+2 facing/unidirectional. **Multiple Working:**

DMSO. Bombardier Derby 2014–15. 22/38.   t.
MSO. Bombardier Derby 2014–15. –/62 1T.   t.
PTSO. Bombardier Derby 2014–15. –/44 1TD 2W.   t.
DMSO. Bombardier Derby 2014–15. –/60.   t.

387 101	**TG**	P		421101	422101	423101	424101
387 102	**TG**	P		421102	422102	423102	424102
387 103	**TG**	P		421103	422103	423103	424103
387 104	**TG**	P		421104	422104	423104	424104
387 105	**TG**	P		421105	422105	423105	424105
387 106	**TG**	P		421106	422106	423106	424106
387 107		P		421107	422107	423107	424107
387 108		P		421108	422108	423108	424108
387 109		P		421109	422109	423109	424109
387 110		P		421110	422110	423110	424110
387 111		P		421111	422111	423111	424111
387 112		P		421112	422112	423112	424112
387 113		P		421113	422113	423113	424113
387 114		P		421114	422114	423114	424114
387 115		P		421115	422115	423115	424115
387 116		P		421116	422116	423116	424116
387 117		P		421117	422117	423117	424117
387 118		P		421118	422118	423118	424118
387 119		P		421119	422119	423119	424119
387 120		P		421120	422120	423120	424120
387 121		P		421121	422121	423121	424121
387 122		P		421122	422122	423122	424122
387 123		P		421123	422123	423123	424123
387 124		P		421124	422124	423124	424124
387 125		P		421125	422125	423125	424125
387 126		P		421126	422126	423126	424126

387 127	P		421127	422127	423127	424127
387 128	P		421128	422128	423128	424128
387 129	P		421129	422129	423129	424129

CLASS 390   PENDOLINO   ALSTOM

Tilting units used on the West Coast Main Line.

Formations: As listed below.
Traction Motors: Two Alstom ONIX 800 of 425 kW.
Wheel Arrangement: 1A-A1 + 1A-A1 + 2-2 + 1A-A1 (+ 2-2 + 1A-A1) + 2-2 + 1A-A1 + 2-2 + 1A-A1 + 1A-A1.
Dimensions: 24.80/23.90 x 2.73 m.
Bogies: Fiat-SIG.
Gangways: Within unit.
Doors: Sliding plug.
Seating Layout: 1: 2+1 facing/unidirectional, 2: 2+2 facing/unidirectional.
Multiple Working: Within class. Can also be controlled from Class 57/3 locos.
Construction: Welded aluminium alloy.
Braking: Disc, rheostatic & regenerative.
Couplers: Dellner 12.
Control System: IGBT Inverter.
Maximum Speed: 125 mph.
Heating & ventilation: Air conditioning.

Units up to 390034 were delivered as 8-car sets, without the TSO (688xx). During 2004–05 these units were increased to 9-cars.

62 extra vehicles were built 2010–12 to lengthen 31 sets to 11-cars. On renumbering units were renumbered by adding 100 to the set number. Four new complete 11-car units were also delivered. All these extra vehicles were built at Savigliano in Italy (all original Pendolino vehicles were built at Birmingham).

390033 was written off in the Lambrigg accident of February 2007.

Advertising livery: 390 104 – Black Alstom vinyls on Virgin silver livery.

DMRFO: Alstom Birmingham/Savigliano 2001–05/2010–12. 18/–. 56.3 t.
MFO(A): Alstom Birmingham/Savigliano 2001–05/2010–12. 37/–(2) 1TD 1W. 52.3 t.
PTFO: Alstom Birmingham/Savigliano 2001–05/2010–12. 44/– 1T. 51.2 t.
MFO(B): Alstom Birmingham/Savigliano 2001–05/2010–12. 46/– 1T. 52.3 t.
(TSO: Alstom Savigliano 2010–12. –/74 1T. 49.2 t.)
(MSO: Alstom Savigliano 2010–12. –/76 1T. 52.2 t.)
TSO: Alstom Birmingham/Savigliano 2001–05/2010–12. –/76 1T. 45.5 t.
MSO(A): Alstom Birmingham/Savigliano 2001–05/2010–12. –/62(4) 1TD 1W. 52.0 t.
PTSRMB: Alstom Birmingham/Savigliano 2001–05/2010–12. –/48. 53.2 t.
MSO(B): Alstom Birmingham/Savigliano 2001–05/2010–12. –/62(2) 1TD 1W. 52.5 t.
DMSO: Alstom Birmingham/Savigliano 2001–05/2010–12. –/46 1T. 54.5 t.

Class 390/0. Original build 9-car units.
Formation: DMRFO–MFO–PTFO–MFO–TSO–MSO–PTSRMB–MSO–DMSO.

390 001	**VT**	A	*VW* MA	69101	69401	69501	69601	68801
				69701	69801	69901	69201	
390 002	**VT**	A	*VW* MA	69102	69402	69502	69602	68802
				69702	69802	69902	69202	
390 005	**VT**	A	*VW* MA	69105	69405	69505	69605	68805
				69705	69805	69905	69205	
390 006	**VT**	A	*VW* MA	69106	69406	69506	69606	68806
				69706	69806	69906	69206	
390 008	**VT**	A	*VW* MA	69108	69408	69508	69608	68808
				69708	69808	69908	69208	

390009	**VT**	A	*VW* MA	69109	69409	69509	69609	68809
				69709	69809	69909	69209	
390010	**VT**	A	*VW* MA	69110	69410	69510	69610	68810
				69710	69810	69910	69210	
390011	**VT**	A	*VW* MA	69111	69411	69511	69611	68811
				69711	69811	69911	69211	
390013	**VT**	A	*VW* MA	69113	69413	69513	69613	68813
				69713	69813	69913	69213	
390016	**VT**	A	*VW* MA	69116	69416	69516	69616	68816
				69716	69816	69916	69216	
390020	**VT**	A	*VW* MA	69120	69420	69520	69620	68820
				69720	69820	69920	69220	
390039	**VT**	A	*VW* MA	69139	69439	69539	69639	68839
				69739	69839	69939	69239	
390040	**VT**	A	*VW* MA	69140	69440	69540	69640	68840
				69740	69840	69940	69240	
390042	**VT**	A	*VW* MA	69142	69442	69542	69642	68842
				69742	69842	69942	69242	
390043	**VT**	A	*VW* MA	69143	69443	69543	69643	68843
				69743	69843	69943	69243	
390044	**VT**	A	*VW* MA	69144	69444	69544	69644	68844
				69744	69844	69944	69244	
390045	**VT**	A	*VW* MA	69145	69445	69545	69645	68845
				69745	69845	69945	69245	
390046	**VT**	A	*VW* MA	69146	69446	69546	69646	68846
				69746	69846	69946	69246	
390047	**VT**	A	*VW* MA	69147	69447	69547	69647	68847
				69747	69847	69947	69247	
390049	**VT**	A	*VW* MA	69149	69449	69549	69649	68849
				69749	69849	69949	69249	
390050	**VT**	A	*VW* MA	69150	69450	69550	69650	68850
				69750	69850	69950	69250	

Class 390/1. Original build 9-car units now extended to 11-cars, except 390 154–157 which were built new as 11-cars.
Formation: DMRFO–MFO–PTFO–MFO–TSO–MSO–TSO–MSO–PTSRMB–MSO–DMSO.

390103	**VT**	A	*VW* MA	69103	69403	69503	69603	65303	68903
				68803	69703	69803	69903	69203	
390104	**AL**	A	*VW* MA	69104	69404	69504	69604	65304	68904
				68804	69704	69804	69904	69204	
390107	**VT**	A	*VW* MA	69107	69407	69507	69607	65307	68907
				68807	69707	69807	69907	69207	
390112	**VT**	A	*VW* MA	69112	69412	69512	69612	65312	68912
				68812	69712	69812	69912	69212	
390114	**VT**	A	*VW* MA	69114	69414	69514	69614	65314	68914
				68814	69714	69814	69914	69214	
390115	**VT**	A	*VW* MA	69115	69415	69515	69615	65315	68915
				68815	69715	69815	69915	69215	
390117	**VT**	A	*VW* MA	69117	69417	69517	69617	65317	68917
				68817	69717	69817	69917	69217	

390 118	**VT**	A	*VW* MA	69118	69418	69518	69618	65318	68918	
				68818	69718	69818	69918	69218		
390 119	**VT**	A	*VW* MA	69119	69419	69519	69619	65319	68919	
				68819	69719	69819	69919	69219		
390 121	**VT**	A	*VW* MA	69121	69421	69521	69621	65321	68921	
				68821	69721	69821	69921	69221		
390 122	**VT**	A	*VW* MA	69122	69422	69522	69622	65322	68922	
				68822	69722	69822	69922	69222		
390 123	**VT**	A	*VW* MA	69123	69423	69523	69623	65323	68923	
				68823	69723	69823	69923	69223		
390 124	**VT**	A	*VW* MA	69124	69424	69524	69624	65324	68924	
				68824	69724	69824	69924	69224		
390 125	**VT**	A	*VW* MA	69125	69425	69525	69625	65325	68925	
				68825	69725	69825	69925	69225		
390 126	**VT**	A	*VW* MA	69126	69426	69526	69626	65326	68926	
				68826	69726	69826	69926	69226		
390 127	**VT**	A	*VW* MA	69127	69427	69527	69627	65327	68927	
				68827	69727	69827	69927	69227		
390 128	**VT**	A	*VW* MA	69128	69428	69528	69628	65328	68928	
				68828	69728	69828	69928	69228		
390 129	**VT**	A	*VW* MA	69129	69429	69529	69629	65329	68929	
				68829	69729	69829	69929	69229		
390 130	**VT**	A	*VW* MA	69130	69430	69530	69630	65330	68930	
				68830	69730	69830	69930	69230		
390 131	**VT**	A	*VW* MA	69131	69431	69531	69631	65331	68931	
				68831	69731	69831	69931	69231		
390 132	**VT**	A	*VW* MA	69132	69432	69532	69632	65332	68932	
				68832	69732	69832	69932	69232		
390 134	**VT**	A	*VW* MA	69134	69434	69534	69634	65334	68934	
				68834	69734	69834	69934	69234		
390 135	**VT**	A	*VW* MA	69135	69435	69535	69635	65335	68935	
				68835	69735	69835	69935	69235		
390 136	**VT**	A	*VW* MA	69136	69436	69536	69636	65336	68936	
				68836	69736	69836	69936	69236		
390 137	**VT**	A	*VW* MA	69137	69437	69537	69637	65337	68937	
				68837	69737	69837	69937	69237		
390 138	**VT**	A	*VW* MA	69138	69438	69538	69638	65338	68938	
				68838	69738	69838	69938	69238		
390 141	**VT**	A	*VW* MA	69141	69441	69541	69641	65341	68941	
				68841	69741	69841	69941	69241		
390 148	**VT**	A	*VW* MA	69148	69448	69548	69648	65348	68948	
				68848	69748	69848	69948	69248		
390 151	**VT**	A	*VW* MA	69151	69451	69551	69651	65351	68951	
				68851	69751	69851	69951	69251		
390 152	**VT**	A	*VW* MA	69152	69452	69552	69652	65352	68952	
				68852	69752	69852	69952	69252		
390 153	**VT**	A	*VW* MA	69153	69453	69553	69653	65353	68953	
				68853	69753	69853	69953	69253		
390 154	**VT**	A	*VW* MA	69154	69454	69554	69654	65354	68954	
				68854	69754	69854	69954	69254		

390 155	**VT**	A	*VW* MA	69155 69455 69555 69655 65355 68955
				68855 69755 69855 69955 69255
390 156	**VT**	A	*VW* MA	69156 69456 69556 69656 65356 68956
				68856 69756 69856 69956 69256
390 157	**VT**	A	*VW* MA	69157 69457 69557 69657 65357 68957
				68857 69757 69857 69957 69257

Names (carried on MFO No. 696xx):

390 001 Virgin Pioneer	390 118 Virgin Princess
390 002 Virgin Angel	390 119 Virgin Warrior
390 005 City of Wolverhampton	390 121 Virgin Dream
390 006 Tate Liverpool	390 122 Penny the Pendolino
390 008 Virgin King	390 123 Virgin Glory
390 009 Treaty of Union	390 124 Virgin Venturer
390 010 A Decade of Progress	390 125 Virgin Stagecoach
390 011 City of Lichfield	390 126 Virgin Enterprise
390 013 Virgin Spirit	390 127 Virgin Buccaneer
390 016 Virgin Champion	390 128 City of Preston
390 020 Virgin Cavalier	390 129 City of Stoke-on-Trent
390 039 Virgin Quest	390 130 City of Edinburgh
390 040 Virgin Pathfinder	390 131 City of Liverpool
390 042 City of Bangor/Dinas Bangor	390 132 City of Birmingham
390 043 Virgin Explorer	390 134 City of Carlisle
390 044 Virgin Lionheart	390 135 City of Lancaster
390 045 101 Squadron	390 136 City of Coventry
390 046 Virgin Soldiers	390 137 Virgin Difference
390 047 CLIC Sargent	390 138 City of London
390 049 Virgin Express	390 141 City of Chester
390 050 Virgin Invader	390 148 Virgin Harrier
390 103 Virgin Hero	390 151 Virgin Ambassador
390 104 Alstom Pendolino	390 152 Virgin Knight
390 107 Virgin Lady	390 153 Mission Accomplished
390 112 Virgin Star	390 154 Matthew Flinders
390 114 City of Manchester	390 155 X-MEN Days of Future Past
390 115 Virgin Crusader	390 156 Stockport 170
390 117 Virgin Prince	390 157 Chad Varah

CLASS 395   HS1 DOMESTIC SETS   HITACHI JAPAN

6-car dual-voltage units used on Southeastern High Speed services from St Pancras to Ashford/Dover/Margate via Ramsgate and Faversham.

Formation: PDTSO–MSO–MSO–MSO–MSO–PDTSO.
Systems: 25 kV AC overhead/750 V DC third rail.
Construction: Aluminium.
Traction Motors: Hitachi asynchronous of 210 kW.
Wheel Arrangement: 2-2 + Bo-Bo + Bo-Bo + Bo-Bo + Bo-Bo + 2-2.
Braking: Disc, rheostatic & regenerative braking.
Dimensions: 20.88/20.0 x 2.81 m.   **Couplers:** Scharfenberg.
Bogies: Hitachi.   **Control System:** IGBT Inverter.
Gangways: Within unit.   **Maximum Speed:** 140 mph.

Doors: Single-leaf sliding.   **Multiple Working:** Within class only.
Heating & ventilation: Air conditioning.
Seating Layout: 2+2 facing/unidirectional (mainly unidirectional).

PDTSO(A): Hitachi Kasado, Japan 2006–09. –/28(12) 1TD 2W. 46.7 t.
MSO: Hitachi Kasado, Japan 2006–09. –/66. 45.0 t–45.7 t.
PDTSO(B): Hitachi Kasado, Japan 2006–09. –/48 1T. 46.7 t.

395 001	**SB**	E	*SE*	AD	39011	39012	39013	39014	39015	39016
395 002	**SB**	E	*SE*	AD	39021	39022	39023	39024	39025	39026
395 003	**SB**	E	*SE*	AD	39031	39032	39033	39034	39035	39036
395 004	**SB**	E	*SE*	AD	39041	39042	39043	39044	39045	39046
395 005	**SB**	E	*SE*	AD	39051	39052	39053	39054	39055	39056
395 006	**SB**	E	*SE*	AD	39061	39062	39063	39064	39065	39066
395 007	**SB**	E	*SE*	AD	39071	39072	39073	39074	39075	39076
395 008	**SB**	E	*SE*	AD	39081	39082	39083	39084	39085	39086
395 009	**SB**	E	*SE*	AD	39091	39092	39093	39094	39095	39096
395 010	**SB**	E	*SE*	AD	39101	39102	39103	39104	39105	39106
395 011	**SB**	E	*SE*	AD	39111	39112	39113	39114	39115	39116
395 012	**SB**	E	*SE*	AD	39121	39122	39123	39124	39125	39126
395 013	**SB**	E	*SE*	AD	39131	39132	39133	39134	39135	39136
395 014	**SB**	E	*SE*	AD	39141	39142	39143	39144	39145	39146
395 015	**SB**	E	*SE*	AD	39151	39152	39153	39154	39155	39156
395 016	**SB**	E	*SE*	AD	39161	39162	39163	39164	39165	39166
395 017	**SB**	E	*SE*	AD	39171	39172	39173	39174	39175	39176
395 018	**SB**	E	*SE*	AD	39181	39182	39183	39184	39185	39186
395 019	**SB**	E	*SE*	AD	39191	39192	39193	39194	39195	39196
395 020	**SB**	E	*SE*	AD	39201	39202	39203	39204	39205	39206
395 021	**SB**	E	*SE*	AD	39211	39212	39213	39214	39215	39216
395 022	**SB**	E	*SE*	AD	39221	39222	39223	39224	39225	39226
395 023	**SB**	E	*SE*	AD	39231	39232	39233	39234	39235	39236
395 024	**SB**	E	*SE*	AD	39241	39242	39243	39244	39245	39246
395 025	**SB**	E	*SE*	AD	39251	39252	39253	39254	39255	39256
395 026	**SB**	E	*SE*	AD	39261	39262	39263	39264	39265	39266
395 027	**SB**	E	*SE*	AD	39271	39272	39273	39274	39275	39276
395 028	**SB**	E	*SE*	AD	39281	39282	39283	39284	39285	39286
395 029	**SB**	E	*SE*	AD	39291	39292	39293	39294	39295	39296

Names (carried on end cars):

395 001	Dame Kelly Holmes	395 018	Mo Farah
395 002	Sebastian Coe	395 019	Jessica Ennis
395 003	Sir Steve Redgrave	395 020	Jason Kenny
395 004	Sir Chris Hoy	395 021	Ed Clancy MBE
395 005	Dame Tanni Grey-Thompson	395 022	Alistair Brownlee
395 006	Daley Thompson	395 023	Ellie Simmonds
395 007	Steve Backley	395 024	Jonnie Peacock
395 008	Ben Ainslie	395 025	Victoria Pendleton
395 009	Rebecca Adlington	395 026	Marc Woods
395 010	Duncan Goodhew	395 027	Hannah Cockcroft
395 011	Katherine Grainger	395 028	Laura Trott
395 016	Jamie Staff	395 029	David Weir
395 017	Dame Sarah Storey		

2. 750 V DC THIRD RAIL EMUs

These classes use the third rail system at 750 V DC (unless stated). Outer couplers are buckeyes on units built before 1982 with bar couplers within the units. Newer units generally have Dellner outer couplers.

CLASS 442   WESSEX EXPRESS   BREL DERBY

Stock built for Waterloo–Bournemouth–Weymouth services. Previously used by South West Trains, all units now used by Southern, principally on Victoria–Gatwick Airport–Brighton services.

Formation: DTSO(A)–TSO–MBC–TSO(W)–DTSO(B).
Construction: Steel.
Traction Motors: Four EE546 of 300 kW recovered from Class 432s.
Wheel Arrangement: 2-2 + 2-2 + Bo-Bo + 2-2 + 2-2.
Braking: Disc.   **Dimensions:** 23.15/23.00 x 2.74 m.
Bogies: Two BREL P7 motor bogies (MBSO). T4 bogies (trailer cars).
Couplers: Buckeye.   **Control System:** 1986-type.
Gangways: Throughout.   **Maximum Speed:** 100 mph.
Doors: Sliding plug.   **Heating & Ventilation:** Air conditioning.
Seating Layout: 1: 2+1 facing, 2: 2+2 mainly unidirectional.
Multiple Working: Within class and Class 33/1 & 73 locos in an emergency.

DTSO(A). Lot No. 31030 Derby 1988–89. –/74. 38.5 t.
TSO. Lot No. 31032 Derby 1988–89. –/76 2T. 37.5 t.
MBC. Lot No. 31034 Derby 1988–89. 24/28. 55.0 t.
TSO(W). Lot No. 31033 Derby 1988–89. –/66(4) 1TD 1T 2W. 37.8 t.
DTSO(B). Lot No. 31031 Derby 1988–89. –/74. 37.3 t.

442 401	GV	A	*SN*	SL	77382	71818	62937	71842	77414
442 402	GV	A	*SN*	SL	77383	71819	62938	71843	77407
442 403	GV	A	*SN*	SL	77384	71820	62941	71844	77408
442 404	GV	A	*SN*	SL	77385	71821	62939	71845	77409
442 405	GV	A	*SN*	SL	77386	71822	62944	71846	77410
442 406	GV	A	*SN*	SL	77389	71823	62942	71847	77411
442 407	GV	A	*SN*	SL	77388	71824	62943	71848	77412
442 408	GV	A	*SN*	SL	77387	71825	62945	71849	77413
442 409	GV	A	*SN*	SL	77390	71826	62946	71850	77406
442 410	GV	A	*SN*	SL	77391	71827	62948	71851	77415
442 411	GV	A	*SN*	SL	77392	71828	62940	71858	77422
442 412	GV	A	*SN*	SL	77393	71829	62947	71853	77417
442 413	GV	A	*SN*	SL	77394	71830	62949	71854	77418
442 414	GV	A	*SN*	SL	77395	71831	62950	71855	77419
442 415	GV	A	*SN*	SL	77396	71832	62951	71856	77420
442 416	GV	A	*SN*	SL	77397	71833	62952	71857	77421
442 417	GV	A	*SN*	SL	77398	71834	62953	71852	77416
442 418	GV	A	*SN*	SL	77399	71835	62954	71859	77423
442 419	GV	A	*SN*	SL	77400	71836	62955	71860	77424
442 420	GV	A	*SN*	SL	77401	71837	62956	71861	77425
442 421	GV	A	*SN*	SL	77402	71838	62957	71862	77426

442 422	**GV**	A	*SN*	SL	77403 71839 62958 71863 77427
442 423	**GV**	A	*SN*	SL	77404 71840 62959 71864 77428
442 424	**GV**	A	*SN*	SL	77405 71841 62960 71865 77429

CLASS 444   DESIRO UK   SIEMENS

Express units.

Formation: DMCO–TSO–TSO–TSORMB–DMSO.
Construction: Aluminium.
Traction Motors: 4 Siemens 1TB2016-0GB02 asynchronous of 250 kW.
Wheel Arrangement: Bo-Bo + 2-2 + 2-2 + 2-2 + Bo-Bo.
Braking: Disc, rheostatic & regenerative. **Dimensions:** 23.57 x 2.80 m.
Bogies: SGP SF5000.   **Couplers:** Dellner 12.
Gangways: Throughout.   **Control System:** IGBT Inverter.
Doors: Single-leaf sliding plug.   **Maximum Speed:** 100 mph.
Heating & Ventilation: Air conditioning.
Seating Layout: 1: 2+1 facing/unidirectional, 2: 2+2 facing/unidirectional.
Multiple Working: Within class and with Class 450.

DMSO. Siemens Vienna/Krefeld 2003–04. –/76. 51.3t.
TSO 67101–145. Siemens Vienna/Krefeld 2003–04. –/76 1T. 40.3t.
TSO 67151–195. Siemens Vienna/Krefeld 2003–04. –/76 1T. 36.8t.
TSORMB. Siemens Vienna/Krefeld 2003–04. –/47 1T 1TD 2W. 42.1t.
DMCO. Siemens Vienna/Krefeld 2003–04. 35/24. 51.3t.

444 001	**ST**	A	*SW*	NT	63801 67101 67151 67201 63851
444 002	**ST**	A	*SW*	NT	63802 67102 67152 67202 63852
444 003	**ST**	A	*SW*	NT	63803 67103 67153 67203 63853
444 004	**ST**	A	*SW*	NT	63804 67104 67154 67204 63854
444 005	**ST**	A	*SW*	NT	63805 67105 67155 67205 63855
444 006	**ST**	A	*SW*	NT	63806 67106 67156 67206 63856
444 007	**ST**	A	*SW*	NT	63807 67107 67157 67207 63857
444 008	**ST**	A	*SW*	NT	63808 67108 67158 67208 63858
444 009	**ST**	A	*SW*	NT	63809 67109 67159 67209 63859
444 010	**ST**	A	*SW*	NT	63810 67110 67160 67210 63860
444 011	**ST**	A	*SW*	NT	63811 67111 67161 67211 63861
444 012	**ST**	A	*SW*	NT	63812 67112 67162 67212 63862
444 013	**ST**	A	*SW*	NT	63813 67113 67163 67213 63863
444 014	**ST**	A	*SW*	NT	63814 67114 67164 67214 63864
444 015	**ST**	A	*SW*	NT	63815 67115 67165 67215 63865
444 016	**ST**	A	*SW*	NT	63816 67116 67166 67216 63866
444 017	**ST**	A	*SW*	NT	63817 67117 67167 67217 63867
444 018	**ST**	A	*SW*	NT	63818 67118 67168 67218 63868
444 019	**ST**	A	*SW*	NT	63819 67119 67169 67219 63869
444 020	**ST**	A	*SW*	NT	63820 67120 67170 67220 63870
444 021	**ST**	A	*SW*	NT	63821 67121 67171 67221 63871
444 022	**ST**	A	*SW*	NT	63822 67122 67172 67222 63872
444 023	**ST**	A	*SW*	NT	63823 67123 67173 67223 63873
444 024	**ST**	A	*SW*	NT	63824 67124 67174 67224 63874
444 025	**ST**	A	*SW*	NT	63825 67125 67175 67225 63875
444 026	**ST**	A	*SW*	NT	63826 67126 67176 67226 63876

444 027	**ST**	A	*SW*	NT	63827 67127 67177 67227 63877
444 028	**ST**	A	*SW*	NT	63828 67128 67178 67228 63878
444 029	**ST**	A	*SW*	NT	63829 67129 67179 67229 63879
444 030	**ST**	A	*SW*	NT	63830 67130 67180 67230 63880
444 031	**ST**	A	*SW*	NT	63831 67131 67181 67231 63881
444 032	**ST**	A	*SW*	NT	63832 67132 67182 67232 63882
444 033	**ST**	A	*SW*	NT	63833 67133 67183 67233 63883
444 034	**ST**	A	*SW*	NT	63834 67134 67184 67234 63884
444 035	**ST**	A	*SW*	NT	63835 67135 67185 67235 63885
444 036	**ST**	A	*SW*	NT	63836 67136 67186 67236 63886
444 037	**ST**	A	*SW*	NT	63837 67137 67187 67237 63887
444 038	**ST**	A	*SW*	NT	63838 67138 67188 67238 63888
444 039	**ST**	A	*SW*	NT	63839 67139 67189 67239 63889
444 040	**ST**	A	*SW*	NT	63840 67140 67190 67240 63890
444 041	**ST**	A	*SW*	NT	63841 67141 67191 67241 63891
444 042	**ST**	A	*SW*	NT	63842 67142 67192 67242 63892
444 043	**ST**	A	*SW*	NT	63843 67143 67193 67243 63893
444 044	**ST**	A	*SW*	NT	63844 67144 67194 67244 63894
444 045	**ST**	A	*SW*	NT	63845 67145 67195 67245 63895

Names (carried on TSORMB):

| 444 001 NAOMI HOUSE | 444 018 THE FAB 444 |
| 444 012 DESTINATION WEYMOUTH | 444 038 SOUTH WESTERN RAILWAY |

CLASS 450　　　　DESIRO UK　　　　SIEMENS

Outer suburban units.

Formation: DMSO–TCO–TSO–DMSO (DMSO–TSO–TCO–DMSO 450 111–127).
Construction: Aluminium.
Traction Motors: 4 Siemens 1TB2016-0GB02 asynchronous of 250 kW.
Wheel Arrangement: Bo-Bo + 2-2 + 2-2 + Bo-Bo.
Braking: Disc, rheostatic & regenerative. **Dimensions:** 20.34 x 2.79 m.
Bogies: SGP SF5000.　　　　　　　　　**Couplers:** Dellner 12.
Gangways: Throughout.　　　　　　　　**Control System:** IGBT Inverter.
Doors: Sliding plug.　　　　　　　　　　**Maximum Speed:** 100 mph.
Heating & Ventilation: Air conditioning.
Seating Layout: 1: 2+2 facing/unidirectional, 2: 3+2 facing/unidirectional.
Multiple Working: Within class and with Class 444.

Class 450/0. Standard units.

DMSO(A). Siemens Krefeld/Vienna 2002–06. –/70. 48.0 t.
TCO. Siemens Krefeld/Vienna 2002–06. 24/32(4) 1T. 35.8 t.
TSO. Siemens Krefeld/Vienna 2002–06. –/61(9) 1TD 2W. 39.8 t.
DMSO(B). Siemens Krefeld/Vienna 2002–06. –/70. 48.6 t.

450 001	**SD**	A	*SW*	NT	63201 64201 68101 63601
450 002	**SD**	A	*SW*	NT	63202 64202 68102 63602
450 003	**SD**	A	*SW*	NT	63203 64203 68103 63603
450 004	**SD**	A	*SW*	NT	63204 64204 68104 63604
450 005	**SD**	A	*SW*	NT	63205 64205 68105 63605
450 006	**SD**	A	*SW*	NT	63206 64206 68106 63606

450007	**SD**	A	*SW*	NT	63207	64207	68107	63607
450008	**SD**	A	*SW*	NT	63208	64208	68108	63608
450009	**SD**	A	*SW*	NT	63209	64209	68109	63609
450010	**SD**	A	*SW*	NT	63210	64210	68110	63610
450011	**SD**	A	*SW*	NT	63211	64211	68111	63611
450012	**SD**	A	*SW*	NT	63212	64212	68112	63612
450013	**SD**	A	*SW*	NT	63213	64213	68113	63613
450014	**SD**	A	*SW*	NT	63214	64214	68114	63614
450015	**SD**	A	*SW*	NT	63215	64215	68115	63615
450016	**SD**	A	*SW*	NT	63216	64216	68116	63616
450017	**SD**	A	*SW*	NT	63217	64217	68117	63617
450018	**SD**	A	*SW*	NT	63218	64218	68118	63618
450019	**SD**	A	*SW*	NT	63219	64219	68119	63619
450020	**SD**	A	*SW*	NT	63220	64220	68120	63620
450021	**SD**	A	*SW*	NT	63221	64221	68121	63621
450022	**SD**	A	*SW*	NT	63222	64222	68122	63622
450023	**SD**	A	*SW*	NT	63223	64223	68123	63623
450024	**SD**	A	*SW*	NT	63224	64224	68124	63624
450025	**SD**	A	*SW*	NT	63225	64225	68125	63625
450026	**SD**	A	*SW*	NT	63226	64226	68126	63626
450027	**SD**	A	*SW*	NT	63227	64227	68127	63627
450028	**SD**	A	*SW*	NT	63228	64228	68128	63628
450029	**SD**	A	*SW*	NT	63229	64229	68129	63629
450030	**SD**	A	*SW*	NT	63230	64230	68130	63630
450031	**SD**	A	*SW*	NT	63231	64231	68131	63631
450032	**SD**	A	*SW*	NT	63232	64232	68132	63632
450033	**SD**	A	*SW*	NT	63233	64233	68133	63633
450034	**SD**	A	*SW*	NT	63234	64234	68134	63634
450035	**SD**	A	*SW*	NT	63235	64235	68135	63635
450036	**SD**	A	*SW*	NT	63236	64236	68136	63636
450037	**SD**	A	*SW*	NT	63237	64237	68137	63637
450038	**SD**	A	*SW*	NT	63238	64238	68138	63638
450039	**SD**	A	*SW*	NT	63239	64239	68139	63639
450040	**SD**	A	*SW*	NT	63240	64240	68140	63640
450041	**SD**	A	*SW*	NT	63241	64241	68141	63641
450042	**SD**	A	*SW*	NT	63242	64242	68142	63642
450071	**SD**	A	*SW*	NT	63271	64271	68171	63671
450072	**SD**	A	*SW*	NT	63272	64272	68172	63672
450073	**SD**	A	*SW*	NT	63273	64273	68173	63673
450074	**SD**	A	*SW*	NT	63274	64274	68174	63674
450075	**SD**	A	*SW*	NT	63275	64275	68175	63675
450076	**SD**	A	*SW*	NT	63276	64276	68176	63676
450077	**SD**	A	*SW*	NT	63277	64277	68177	63677
450078	**SD**	A	*SW*	NT	63278	64278	68178	63678
450079	**SD**	A	*SW*	NT	63279	64279	68179	63679
450080	**SD**	A	*SW*	NT	63280	64280	68180	63680
450081	**SD**	A	*SW*	NT	63281	64281	68181	63681
450082	**SD**	A	*SW*	NT	63282	64282	68182	63682
450083	**SD**	A	*SW*	NT	63283	64283	68183	63683
450084	**SD**	A	*SW*	NT	63284	64284	68184	63684
450085	**SD**	A	*SW*	NT	63285	64285	68185	63685

450 086	**SD**	A	*SW*	NT	63286	64286	68186	63686
450 087	**SD**	A	*SW*	NT	63287	64287	68187	63687
450 088	**SD**	A	*SW*	NT	63288	64288	68188	63688
450 089	**SD**	A	*SW*	NT	63289	64289	68189	63689
450 090	**SD**	A	*SW*	NT	63290	64290	68190	63690
450 091	**SD**	A	*SW*	NT	63291	64291	68191	63691
450 092	**SD**	A	*SW*	NT	63292	64292	68192	63692
450 093	**SD**	A	*SW*	NT	63293	64293	68193	63693
450 094	**SD**	A	*SW*	NT	63294	64294	68194	63694
450 095	**SD**	A	*SW*	NT	63295	64295	68195	63695
450 096	**SD**	A	*SW*	NT	63296	64296	68196	63696
450 097	**SD**	A	*SW*	NT	63297	64297	68197	63697
450 098	**SD**	A	*SW*	NT	63298	64298	68198	63698
450 099	**SD**	A	*SW*	NT	63299	64299	68199	63699
450 100	**SD**	A	*SW*	NT	63300	64300	68200	63700
450 101	**SD**	A	*SW*	NT	63701	66851	66801	63751
450 102	**SD**	A	*SW*	NT	63702	66852	66802	63752
450 103	**SD**	A	*SW*	NT	63703	66853	66803	63753
450 104	**SD**	A	*SW*	NT	63704	66854	66804	63754
450 105	**SD**	A	*SW*	NT	63705	66855	66805	63755
450 106	**SD**	A	*SW*	NT	63706	66856	66806	63756
450 107	**SD**	A	*SW*	NT	63707	66857	66807	63757
450 108	**SD**	A	*SW*	NT	63708	66858	66808	63758
450 109	**SD**	A	*SW*	NT	63709	66859	66809	63759
450 110	**SD**	A	*SW*	NT	63710	66860	66810	63760
450 111	**SD**	A	*SW*	NT	63901	66921	66901	63921
450 112	**SD**	A	*SW*	NT	63902	66922	66902	63922
450 113	**SD**	A	*SW*	NT	63903	66923	66903	63923
450 114	**SD**	A	*SW*	NT	63904	66924	66904	63924
450 115	**SD**	A	*SW*	NT	63905	66925	66905	63925
450 116	**SD**	A	*SW*	NT	63906	66926	66906	63926
450 117	**SD**	A	*SW*	NT	63907	66927	66907	63927
450 118	**SD**	A	*SW*	NT	63908	66928	66908	63928
450 119	**SD**	A	*SW*	NT	63909	66929	66909	63929
450 120	**SD**	A	*SW*	NT	63910	66930	66910	63930
450 121	**SD**	A	*SW*	NT	63911	66931	66911	63931
450 122	**SD**	A	*SW*	NT	63912	66932	66912	63932
450 123	**SD**	A	*SW*	NT	63913	66933	66913	63933
450 124	**SD**	A	*SW*	NT	63914	66934	66914	63934
450 125	**SD**	A	*SW*	NT	63915	66935	66915	63935
450 126	**SD**	A	*SW*	NT	63916	66936	66916	63936
450 127	**SD**	A	*SW*	NT	63917	66937	66917	63937

Names (carried on DMSO(B)):

450 015 DESIRO
450 042 TRELOAR COLLEGE

450 114 FAIRBRIDGE investing in the future

Class 450/5. 28 units converted 2007–08 with First Class removed and a modified seating layout with more standing room (some Standard Class seats were taken out). First Class was refitted in 2013 but the removed Standard Class seats were not refitted so the units have kept their 450 5xx series numbers.

DMSO(A). Siemens Krefeld/Vienna 2002–04. –/64. 48.0 t.
TCO. Siemens Krefeld/Vienna 2002–04. 24/30(4) 1T. 35.5 t.
TSO. Siemens Krefeld/Vienna 2002–04. –/56(9) 1TD 2W. 39.8 t.
DMSO(B). Siemens Krefeld/Vienna 2002–04. –/64. 48.6 t.

450543	(450043)	**SD**	A	*SW*	NT	63243	64243	68143	63643
450544	(450044)	**SD**	A	*SW*	NT	63244	64244	68144	63644
450545	(450045)	**SD**	A	*SW*	NT	63245	64245	68145	63645
450546	(450046)	**SD**	A	*SW*	NT	63246	64246	68146	63646
450547	(450047)	**SD**	A	*SW*	NT	63247	64247	68147	63647
450548	(450048)	**SD**	A	*SW*	NT	63248	64248	68148	63648
450549	(450049)	**SD**	A	*SW*	NT	63249	64249	68149	63649
450550	(450050)	**SD**	A	*SW*	NT	63250	64250	68150	63650
450551	(450051)	**SD**	A	*SW*	NT	63251	64251	68151	63651
450552	(450052)	**SD**	A	*SW*	NT	63252	64252	68152	63652
450553	(450053)	**SD**	A	*SW*	NT	63253	64253	68153	63653
450554	(450054)	**SD**	A	*SW*	NT	63254	64254	68154	63654
450555	(450055)	**SD**	A	*SW*	NT	63255	64255	68155	63655
450556	(450056)	**SD**	A	*SW*	NT	63256	64256	68156	63656
450557	(450057)	**SD**	A	*SW*	NT	63257	64257	68157	63657
450558	(450058)	**SD**	A	*SW*	NT	63258	64258	68158	63658
450559	(450059)	**SD**	A	*SW*	NT	63259	64259	68159	63659
450560	(450060)	**SD**	A	*SW*	NT	63260	64260	68160	63660
450561	(450061)	**SD**	A	*SW*	NT	63261	64261	68161	63661
450562	(450062)	**SD**	A	*SW*	NT	63262	64262	68162	63662
450563	(450063)	**SD**	A	*SW*	NT	63263	64263	68163	63663
450564	(450064)	**SD**	A	*SW*	NT	63264	64264	68164	63664
450565	(450065)	**SD**	A	*SW*	NT	63265	64265	68165	63665
450566	(450066)	**SD**	A	*SW*	NT	63266	64266	68166	63666
450567	(450067)	**SD**	A	*SW*	NT	63267	64267	68167	63667
450568	(450068)	**SD**	A	*SW*	NT	63268	64268	68168	63668
450569	(450069)	**SD**	A	*SW*	NT	63269	64269	68169	63669
450570	(450070)	**SD**	A	*SW*	NT	63270	64270	68170	63670

CLASS 455   BREL YORK

Inner suburban units.

Formation: DTSO–MSO–TSO–DTSO.
Construction: Steel. Class 455/7 TSO have a steel underframe and an aluminium alloy body & roof.
Traction Motors: Four GEC507-20J of 185 kW, some recovered from Class 405s.
Wheel Arrangement: 2-2 + Bo-Bo + 2-2 + 2-2.
Braking: Disc.   **Dimensions:** 19.92/19.83 x 2.82 m.
Bogies: P7 (motor) and T3 (455/8 & 455/7) BX1 (455/7) trailer.
Gangways: Within unit + end doors (sealed on Southern units).
Couplers: Tightlock.   **Control System:** 1982-type, camshaft.
Doors: Sliding.   **Maximum Speed:** 75 mph.
Heating & Ventilation: Various.
Seating Layout: All units refurbished. SWT units: 2+2 high-back unidirectional/facing seating. Southern units: 3+2 high back mainly facing seating.
Multiple Working: Within class and with Class 456.

Class 455/7. South West Trains units. Second series with TSOs originally in
Class 508s. Pressure heating & ventilation.

DTSO. Lot No. 30976 1984–85. –/50(4) 1W. 30.8t.
MSO. Lot No. 30975 1984–85. –/68. 45.7t.
TSO. Lot No. 30944 1979–80. –/68. 26.1t.

5701	**SS**	P	*SW*	WD	77727	62783	71545	77728
5702	**SS**	P	*SW*	WD	77729	62784	71547	77730
5703	**SS**	P	*SW*	WD	77731	62785	71540	77732
5704	**SS**	P	*SW*	WD	77733	62786	71548	77734
5705	**SS**	P	*SW*	WD	77735	62787	71565	77736
5706	**SS**	P	*SW*	WD	77737	62788	71534	77738
5707	**SS**	P	*SW*	WD	77739	62789	71536	77740
5708	**SS**	P	*SW*	WD	77741	62790	71560	77742
5709	**SS**	P	*SW*	WD	77743	62791	71532	77744
5710	**SS**	P	*SW*	WD	77745	62792	71566	77746
5711	**SS**	P	*SW*	WD	77747	62793	71542	77748
5712	**SS**	P	*SW*	WD	77749	62794	71546	77750
5713	**SS**	P	*SW*	WD	77751	62795	71567	77752
5714	**SS**	P	*SW*	WD	77753	62796	71539	77754
5715	**SS**	P	*SW*	WD	77755	62797	71535	77756
5716	**SS**	P	*SW*	WD	77757	62798	71564	77758
5717	**SS**	P	*SW*	WD	77759	62799	71528	77760
5718	**SS**	P	*SW*	WD	77761	62800	71557	77762
5719	**SS**	P	*SW*	WD	77763	62801	71558	77764
5720	**SS**	P	*SW*	WD	77765	62802	71568	77766
5721	**SS**	P	*SW*	WD	77767	62803	71553	77768
5722	**SS**	P	*SW*	WD	77769	62804	71533	77770
5723	**SS**	P	*SW*	WD	77771	62805	71526	77772
5724	**SS**	P	*SW*	WD	77773	62806	71561	77774
5725	**SS**	P	*SW*	WD	77775	62807	71541	77776
5726	**SS**	P	*SW*	WD	77777	62808	71556	77778
5727	**SS**	P	*SW*	WD	77779	62809	71562	77780
5728	**SS**	P	*SW*	WD	77781	62810	71527	77782
5729	**SS**	P	*SW*	WD	77783	62811	71550	77784
5730	**SS**	P	*SW*	WD	77785	62812	71551	77786
5731	**SS**	P	*SW*	WD	77787	62813	71555	77788
5732	**SS**	P	*SW*	WD	77789	62814	71552	77790
5733	**SS**	P	*SW*	WD	77791	62815	71549	77792
5734	**SS**	P	*SW*	WD	77793	62816	71531	77794
5735	**SS**	P	*SW*	WD	77795	62817	71563	77796
5736	**SS**	P	*SW*	WD	77797	62818	71554	77798
5737	**SS**	P	*SW*	WD	77799	62819	71544	77800
5738	**SS**	P	*SW*	WD	77801	62820	71529	77802
5739	**SS**	P	*SW*	WD	77803	62821	71537	77804
5740	**SS**	P	*SW*	WD	77805	62822	71530	77806
5741	**SS**	P	*SW*	WD	77807	62823	71559	77808
5742	**SS**	P	*SW*	WD	77809	62824	71543	77810
5750	**SS**	P	*SW*	WD	77811	62825	71538	77812

Class 455/8. Southern units. First series. Pressure heating & ventilation. Fitted with in-cab air conditioning systems meaning that the end door has been sealed.

DTSO. Lot No. 30972 York 1982–84. –/74. 33.6 t.
MSO. Lot No. 30973 York 1982–84. –/84. 45.6 t.
TSO. Lot No. 30974 York 1982–84. –/75(3) 2W. 34.0 t.

455801	**SN**	E	*SN*	SU	77627	62709	71657	77580
455802	**SN**	E	*SN*	SU	77581	62710	71664	77582
455803	**SN**	E	*SN*	SU	77583	62711	71639	77584
455804	**SN**	E	*SN*	SU	77585	62712	71640	77586
455805	**SN**	E	*SN*	SU	77587	62713	71641	77588
455806	**SN**	E	*SN*	SU	77589	62714	71642	77590
455807	**SN**	E	*SN*	SU	77591	62715	71643	77592
455808	**SN**	E	*SN*	SU	77637	62716	71644	77594
455809	**SN**	E	*SN*	SU	77623	62717	71648	77602
455810	**SN**	E	*SN*	SU	77597	62718	71646	77598
455811	**SN**	E	*SN*	SU	77599	62719	71647	77600
455812	**SN**	E	*SN*	SU	77595	62720	71645	77626
455813	**SN**	E	*SN*	SU	77603	62721	71649	77604
455814	**SN**	E	*SN*	SU	77605	62722	71650	77606
455815	**SN**	E	*SN*	SU	77607	62723	71651	77608
455816	**SN**	E	*SN*	SU	77609	62724	71652	77633
455817	**SN**	E	*SN*	SU	77611	62725	71653	77612
455818	**SN**	E	*SN*	SU	77613	62726	71654	77632
455819	**SN**	E	*SN*	SU	77615	62727	71637	77616
455820	**SN**	E	*SN*	SU	77617	62728	71656	77618
455821	**SN**	E	*SN*	SU	77619	62729	71655	77620
455822	**SN**	E	*SN*	SU	77621	62730	71658	77622
455823	**SN**	E	*SN*	SU	77601	62731	71659	77596
455824	**SN**	E	*SN*	SU	77593	62732	71660	77624
455825	**SN**	E	*SN*	SU	77579	62733	71661	77628
455826	**SN**	E	*SN*	SU	77630	62734	71662	77629
455827	**SN**	E	*SN*	SU	77610	62735	71663	77614
455828	**SN**	E	*SN*	SU	77631	62736	71638	77634
455829	**SN**	E	*SN*	SU	77635	62737	71665	77636
455830	**SN**	E	*SN*	SU	77625	62743	71666	77638
455831	**SN**	E	*SN*	SU	77639	62739	71667	77640
455832	**SN**	E	*SN*	SU	77641	62740	71668	77642
455833	**SN**	E	*SN*	SU	77643	62741	71669	77644
455834	**SN**	E	*SN*	SU	77645	62742	71670	77646
455835	**SN**	E	*SN*	SU	77647	62738	71671	77648
455836	**SN**	E	*SN*	SU	77649	62744	71672	77650
455837	**SN**	E	*SN*	SU	77651	62745	71673	77652
455838	**SN**	E	*SN*	SU	77653	62746	71674	77654
455839	**SN**	E	*SN*	SU	77655	62747	71675	77656
455840	**SN**	E	*SN*	SU	77657	62748	71676	77658
455841	**SN**	E	*SN*	SU	77659	62749	71677	77660
455842	**SN**	E	*SN*	SU	77661	62750	71678	77662
455843	**SN**	E	*SN*	SU	77663	62751	71679	77664
455844	**SN**	E	*SN*	SU	77665	62752	71680	77666

| 455845 | **SN** | E | *SN* | SU | 77667 | 62753 | 71681 | 77668 |
| 455846 | **SN** | E | *SN* | SU | 77669 | 62754 | 71682 | 77670 |

Class 455/8. South West Trains units. First series. Pressure heating & ventilation.

DTSO. Lot No. 30972 York 1982–84. –50(4) 1W. 29.5 t.
MSO. Lot No. 30973 York 1982–84. –/84 –/68. 45.6 t.
TSO. Lot No. 30974 York 1982–84. –/84 –/68. 27.1 t.

5847	**SS**	P	*SW*	WD	77671	62755	71683	77672
5848	**SS**	P	*SW*	WD	77673	62756	71684	77674
5849	**SS**	P	*SW*	WD	77675	62757	71685	77676
5850	**SS**	P	*SW*	WD	77677	62758	71686	77678
5851	**SS**	P	*SW*	WD	77679	62759	71687	77680
5852	**SS**	P	*SW*	WD	77681	62760	71688	77682
5853	**SS**	P	*SW*	WD	77683	62761	71689	77684
5854	**SS**	P	*SW*	WD	77685	62762	71690	77686
5855	**SS**	P	*SW*	WD	77687	62763	71691	77688
5856	**SS**	P	*SW*	WD	77689	62764	71692	77690
5857	**SS**	P	*SW*	WD	77691	62765	71693	77692
5858	**SS**	P	*SW*	WD	77693	62766	71694	77694
5859	**SS**	P	*SW*	WD	77695	62767	71695	77696
5860	**SS**	P	*SW*	WD	77697	62768	71696	77698
5861	**SS**	P	*SW*	WD	77699	62769	71697	77700
5862	**SS**	P	*SW*	WD	77701	62770	71698	77702
5863	**SS**	P	*SW*	WD	77703	62771	71699	77704
5864	**SS**	P	*SW*	WD	77705	62772	71700	77706
5865	**SS**	P	*SW*	WD	77707	62773	71701	77708
5866	**SS**	P	*SW*	WD	77709	62774	71702	77710
5867	**SS**	P	*SW*	WD	77711	62775	71703	77712
5868	**SS**	?	*SW*	WD	77713	62776	71704	77714
5869	**SS**	P	*SW*	WD	77715	62777	71705	77716
5870	**SS**	P	*SW*	WD	77717	62778	71706	77718
5871	**SS**	P	*SW*	WD	77719	62779	71707	77720
5872	**SS**	P	*SW*	WD	77721	62780	71708	77722
5873	**SS**	P	*SW*	WD	77723	62781	71709	77724
5874	**SS**	P	*SW*	WD	77725	62782	71710	77726

Class 455/9. South West Trains units. Third series. Convection heating.
Dimensions: 19.96/20.18 x 2.82 m.

Vehicles 67301 and 67400 were converted from Class 210 DEMU vehicles to replace accident damaged cars.

DTSO. Lot No. 30991 York 1985. –/50(4) 1W. 30.7 t.
MSO. Lot No. 30992 York 1985. –/68. 46.3 t.
TSO. Lot No. 30993 York 1985. –/68. 28.3 t.
TSO†. Lot No. 30932 Derby 1981. –/68. 26.5 t.

5901	**SS**	P	*SW*	WD	77813	62826	71714	77814
5902	**SS**	P	*SW*	WD	77815	62827	71715	77816
5903	**SS**	P	*SW*	WD	77817	62828	71716	77818
5904	**SS**	P	*SW*	WD	77819	62829	71717	77820
5905	**SS**	P	*SW*	WD	77821	62830	71725	77822

5906		**SS**	P	*SW*	WD	77823	62831	71719	77824
5907		**SS**	P	*SW*	WD	77825	62832	71720	77826
5908		**SS**	P	*SW*	WD	77827	62833	71721	77828
5909		**SS**	P	*SW*	WD	77829	62834	71722	77830
5910		**SS**	P	*SW*	WD	77831	62835	71723	77832
5911		**SS**	P	*SW*	WD	77833	62836	71724	77834
5912	†	**SS**	P	*SW*	WD	77835	62837	67400	77836
5913	†	**SS**	P	*SW*	WD	77837	67301	71726	77838
5914		**SS**	P	*SW*	WD	77839	62839	71727	77840
5915		**SS**	P	*SW*	WD	77841	62840	71728	77842
5916		**SS**	P	*SW*	WD	77843	62841	71729	77844
5917		**SS**	P	*SW*	WD	77845	62842	71730	77846
5918		**SS**	P	*SW*	WD	77847	62843	71732	77848
5919		**SS**	P	*SW*	WD	77849	62844	71718	77850
5920		**SS**	P	*SW*	WD	77851	62845	71733	77852

CLASS 456   BREL YORK

Inner suburban units. These units were withdrawn from traffic with Southern in late 2013.

During 2014–15 all units will be refurbished at Wolverton for South West Trains. Whilst this takes place some units are in traffic with SWT in unrefurbished condition.

Formation: DMSO–DTSO.
Construction: Steel underframe, aluminium alloy body & roof.
Traction Motors: Two GEC507-20J of 185 kW, some recovered from Class 405s.
Wheel Arrangement: 2-Bo + 2-2.   **Dimensions:** 20.61 x 2.82 m.
Braking: Disc.   **Couplers:** Tightlock.
Bogies: P7 (motor) and T3 (trailer).   **Control System:** GTO Chopper.
Gangways: Within unit.   **Maximum Speed:** 75 mph.
Doors: Sliding.
Seating Layout: 3+2 facing (* 2+2 facing/unidirectional).
Heating & Ventilation: Convection heating.
Multiple Working: Within class and with Class 455.

Advertising livery: 456 006 TfL/City of London (blue & green with various images).

DMSO. Lot No. 31073 1990–91. –/79 (* –/59). 41.1t (* 43.3 t).
DTSO. Lot No. 31074 1990–91. –/73 (* –/54(5)). 31.4t (* 32.3 t).

456 001		**SN**	P	*SW*	WD	64735	78250
456 002		**SN**	P		ZN	64736	78251
456 003	*	**SS**	P	*SW*	WD	64737	78252
456 004		**SN**	P		WD	64738	78253
456 005		**SN**	P	*SW*	WD	64739	78254
456 006		**AL**	P		ZN	64740	78255
456 007		**SN**	P	*SW*	WD	64741	78256
456 008		**SN**	P	*SW*	WD	64742	78257
456 009		**SN**	P	*SW*	WD	64743	78258

456 010	**SN**	P		ZN	64744	78259
456 011	**SN**	P	*SW*	WD	64745	78260
456 012	**SN**	P		ZN	64746	78261
456 013	**SN**	P	*SW*	WD	64747	78262
456 014	* **SS**	P	*SW*	WD	64748	78263
456 015	**SN**	P		ZN	64749	78264
456 016	**SN**	P	*SW*	WD	64750	78265
456 017	**SN**	P	*SW*	WD	64751	78266
456 018	**SN**	P		WD	64752	78267
456 019	**SN**	P	*SW*	WD	64753	78268
456 020	**SN**	P	*SW*	WD	64754	78269
456 021	**SN**	P	*SW*	WD	64755	78270
456 022	**SN**	P	*SW*	WD	64756	78271
456 023	**SN**	P	*SW*	WD	64757	78272
456 024	**SN**	P		ZN	64758	78273

CLASS 458   JUNIPER   ALSTOM BIRMINGHAM

Outer suburban units. Between 2013 and 2015 the fleet of 30 4-car Class 458 units and the former Gatwick Express fleet of eight 8-car Class 460 units is being combined to form a fleet of 36 5-car Standard Class only Class 458/5s. The work is being carried out at Wabtec Doncaster and Brush Loughborough. Former Class 460 driving cars 67901/903/907/908 are not included in this programme and are shown in the EMUs Awaiting Disposal section of this book.

The first six units to be completed were 458 531–536, which use all former Class 460 vehicles. These are being followed by 8001–30 which are being augmented to 5-cars. All of the spare 460 vehicles to be added to 458s are listed at the end of this section, with a space left in the formations of 8001–30 for them to be inserted.

After lengthening each unit is being renumbered into the 458 5xx series. All individual vehicles retain their original numbers.

Formation: DMCO–TSO–MSO–DMCO (as built) or 5-cars DMSO–TSO–TSO–MSO–DMSO.
Construction: Steel.   **Dimensions:** 21.16 or 21.06 x 2.80 m.
Traction Motors: Two Alstom ONIX 800 asynchronous of 270 kW.
Wheel Arrangement: 2-Bo + 2-2 (+ 2-2) + Bo-2 + Bo-2.
Braking: Disc & regenerative.   **Control System:** IGBT Inverter.
Bogies: ACR.   **Doors:** Sliding plug.
Gangways: Throughout.
Couplers: Scharfenberg AAR (458/5 Voith 136).
Maximum Speed: 100 mph (458/5 units being regeared to 75 mph).
Heating & Ventilation: Air conditioning. **Multiple Working:** Within class.
Seating Layout: 1: 2+2 facing, 2: 3+2 facing/unidirectional (458/5 units Standard Class only, 2+2 facing/unidirectional).

DMCO(A). Alstom 1998–2000. 12/63 (458/5 –/60). 46.4 t/45.7 t.
TSO (ex-460): Alstom 1998–99. 458 531–536 –/52 1T (for 8001–30 –/56). 34.4 t.
TSO. Alstom 1998–2000. –/54(6) 1TD 2W (458/5 –/42 1TD 2W). 34.6 t/34.1 t.
MSO. Alstom 1998–2000. –/75 1T (458 531–536 –/56, 458 501–530 –56 1T).
42.1 t/39.3 t.
DMCO(B). Alstom 1998–2000. 12/63 (458/5 –/60). 46.4 t/45.7 t.

458501	**SD**	P	*SW*	WD	67601	74431	74001	74101	67701
458502	**SD**	P	*SW*	WD	67602	74421	74002	74102	67702
458503	**SD**	P	*SW*	WD	67603	74441	74003	74103	67703
8004	**ST**	P	*SW*	WD	67604		74004	74104	67704
8005	**ST**	P	*SW*	WD	67605		74005	74105	67705
8006	**ST**	P	*SW*	WD	67606		74006	74106	67706
8007	**ST**	P	*SW*	WD	67607		74007	74107	67707
8008	**ST**	P	*SW*	WD	67608		74008	74108	67708
8009	**ST**	P	*SW*	WD	67609		74009	74109	67709
8010	**ST**	P	*SW*	WD	67610		74010	74110	67710
8011	**ST**	P	*SW*	WD	67611		74011	74111	67711
8012	**ST**	P	*SW*	WD	67612		74012	74112	67712
8013	**ST**	P	*SW*	WD	67613		74013	74113	67713
8014	**ST**	P	*SW*	WD	67614		74014	74114	67714
8015	**ST**	P	*SW*	WD	67615		74015	74115	67715
8016	**ST**	P	*SW*	WD	67616		74016	74116	67716
8017	**ST**	P	*SW*	WD	67617		74017	74117	67717
8018	**ST**	P	*SW*	WD	67618		74018	74118	67718
8019	**ST**	P	*SW*	WD	67619		74019	74119	67719
8020	**ST**	P	*SW*	WD	67620		74020	74120	67720
8021	**ST**	P	*SW*	WD	67621		74021	74121	67721
8022	**ST**	P	*SW*	WD	67622		74022	74122	67722
8023	**ST**	P	*SW*	WD	67623		74023	74123	67723
8024	**ST**	P	*SW*	WD	67624		74024	74124	67724
8025	**ST**	P	*SW*	WD	67625		74025	74125	67725
8026	**ST**	P	*SW*	WD	67626		74026	74126	67726
8027	**ST**	P	*SW*	WD	67627		74027	74127	67727
8028	**ST**	P	*SW*	WD	67628		74028	74128	67728
8029	**ST**	P	*SW*	WD	67629		74029	74129	67729
8030	**ST**	P	*SW*	WD	67630		74030	74130	67730

The following Class 460 vehicles will be added to sets 8004–30 during 2014–15.

74401	74402	74403	74404	74405	74406	74407	74408	74411
74412	74422	74423	74424	74425	74426	74427	74428	74432
74433	74434	74435	74436	74437	74438	74442	74451	74452

The following units have been converted from Class 460s.

458531	**SD**	P	*SW*	WD	67913	74418	74446	74458	67912
458532	**SD**	P	*SW*	WD	67904	74417	74447	74457	67905
458533	**SD**	P	*SW*	WD	67917	74413	74443	74453	67916
458534	**SD**	P	*SW*	WD	67914	74414	74444	74454	67918
458535	**SD**	P	*SW*	WD	67915	74415	74445	74455	67911
458536	**SD**	P	*SW*	WD	67906	74416	74448	74456	67902

CLASS 465   NETWORKER

Inner/outer suburban units.

Formation: DMSO–TSO–TSO–DMSO.
Construction: Welded aluminium alloy.
Traction Motors: Hitachi asynchronous of 280 kW (Classes 465/0 and 465/1) or GEC-Alsthom G352BY (Classes 465/2 and 465/9).
Wheel Arrangement: Bo-Bo + 2-2 + 2-2 + Bo-Bo.
Braking: Disc & rheostatic and regenerative (Classes 465/0 and 465/1 only).
Bogies: BREL P3/T3 (465/0 and 465/1), SRP BP62/BT52 (465/2 and 465/9).
Dimensions: 20.89/20.06 x 2.81 m.
Control System: IGBT Inverter (465/0 and 465/1) or 1992-type GTO Inverter.
Gangways: Within unit.   **Couplers:** Tightlock.
Doors: Sliding plug.   **Maximum Speed:** 75 mph.
Seating Layout: 3+2 facing/unidirectional.
Multiple Working: Within class and with Class 466.

64759–808. DMSO(A). Lot No. 31100 BREL York 1991–93. –/86. 39.2t.
64809–858. DMSO(B). Lot No. 31100 BREL York 1991–93. –/86. 39.2t.
65734–749. DMSO(A). Lot No. 31103 Metro-Cammell 1991–93. –/86. 39.2t.
65784–799. DMSO(B). Lot No. 31103 Metro-Cammell 1991–93. –/86. 39.2t.
65800–846. DMSO(A). Lot No. 31130 ABB York 1993–94. –/86. 39.2t.
65847–893. DMSO(B). Lot No. 31130 ABB York 1993–94. –/86. 39.2t.
72028–126 (even nos.) TSO. Lot No. 31102 BREL York 1991–93. –/90. 27.2t.
72029–127 (odd nos.) TSO. Lot No. 31101 BREL York 1991–93. –/86 1T. 28.0t.
72787–817 (odd nos.) TSO. Lot No. 31104 Metro-Cammell 1991–92. –/86 1T. 28.0t.
72788–818 (even nos.) TSO. Lot No. 31105 Metro-Cammell 1991–92. –/90. 27.2t.
72900–992 (even nos.) TSO. Lot No. 31102 ABB York 1993–94. –/90. 27.2t.
72901–993 (odd nos.) TSO. Lot No. 31101 ABB York 1993–94. –/86 1T. 28.0t.

Class 465/0. Built by BREL/ABB.

465001	**SE**	E	*SE*	SG	64759	72028	72029	64809
465002	**SE**	E	*SE*	SG	64760	72030	72031	64810
465003	**SE**	E	*SE*	SG	64761	72032	72033	64811
465004	**SE**	E	*SE*	SG	64762	72034	72035	64812
465005	**SE**	E	*SE*	SG	64763	72036	72037	64813
465006	**SE**	E	*SE*	SG	64764	72038	72039	64814
465007	**SE**	E	*SE*	SG	64765	72040	72041	64815
465008	**SE**	E	*SE*	SG	64766	72042	72043	64816
465009	**SE**	E	*SE*	SG	64767	72044	72045	64817
465010	**SE**	E	*SE*	SG	64768	72046	72047	64818
465011	**SE**	E	*SE*	SG	64769	72048	72049	64819
465012	**SE**	E	*SE*	SG	64770	72050	72051	64820
465013	**SE**	E	*SE*	SG	64771	72052	72053	64821
465014	**SE**	E	*SE*	SG	64772	72054	72055	64822
465015	**SE**	E	*SE*	SG	64773	72056	72057	64823
465016	**SE**	E	*SE*	SG	64774	72058	72059	64824
465017	**SE**	E	*SE*	SG	64775	72060	72061	64825
465018	**SE**	E	*SE*	SG	64776	72062	72063	64826
465019	**SE**	E	*SE*	SG	64777	72064	72065	64827

465 020	**SE**	E	*SE*	SG	64778	72066	72067	64828
465 021	**SE**	E	*SE*	SG	64779	72068	72069	64829
465 022	**SE**	E	*SE*	SG	64780	72070	72071	64830
465 023	**SE**	E	*SE*	SG	64781	72072	72073	64831
465 024	**SE**	E	*SE*	SG	64782	72074	72075	64832
465 025	**SE**	E	*SE*	SG	64783	72076	72077	64833
465 026	**SE**	E	*SE*	SG	64784	72078	72079	64834
465 027	**SE**	E	*SE*	SG	64785	72080	72081	64835
465 028	**SE**	E	*SE*	SG	64786	72082	72083	64836
465 029	**SE**	E	*SE*	SG	64787	72084	72085	64837
465 030	**SE**	E	*SE*	SG	64788	72086	72087	64838
465 031	**SE**	E	*SE*	SG	64789	72088	72089	64839
465 032	**SE**	E	*SE*	SG	64790	72090	72091	64840
465 033	**SE**	E	*SE*	SG	64791	72092	72093	64841
465 034	**SE**	E	*SE*	SG	64792	72094	72095	64842
465 035	**SE**	E	*SE*	SG	64793	72096	72097	64843
465 036	**SE**	E	*SE*	SG	64794	72098	72099	64844
465 037	**SE**	E	*SE*	SG	64795	72100	72101	64845
465 038	**SE**	E	*SE*	SG	64796	72102	72103	64846
465 039	**SE**	E	*SE*	SG	64797	72104	72105	64847
465 040	**SE**	E	*SE*	SG	64798	72106	72107	64848
465 041	**SE**	E	*SE*	SG	64799	72108	72109	64849
465 042	**SE**	E	*SE*	SG	64800	72110	72111	64850
465 043	**SE**	E	*SE*	SG	64801	72112	72113	64851
465 044	**SE**	E	*SE*	SG	64802	72114	72115	64852
465 045	**SE**	E	*SE*	SG	64803	72116	72117	64853
465 046	**SE**	E	*SE*	SG	64804	72118	72119	64854
465 047	**SE**	E	*SE*	SG	64805	72120	72121	64855
465 048	**SE**	E	*SE*	SG	64806	72122	72123	64856
465 049	**SE**	E	*SE*	SG	64807	72124	72125	64857
465 050	**SE**	E	*SE*	SG	64808	72126	72127	64858

Class 465/1. Built by BREL/ABB. Similar to Class 465/0 but with detail differences.

465 151	**SE**	E	*SE*	SG	65800	72900	72901	65847
465 152	**SE**	E	*SE*	SG	65801	72902	72903	65848
465 153	**SE**	E	*SE*	SG	65802	72904	72905	65849
465 154	**SE**	E	*SE*	SG	65803	72906	72907	65850
465 155	**SE**	E	*SE*	SG	65804	72908	72909	65851
465 156	**SE**	E	*SE*	SG	65805	72910	72911	65852
465 157	**SE**	E	*SE*	SG	65806	72912	72913	65853
465 158	**SE**	E	*SE*	SG	65807	72914	72915	65854
465 159	**SE**	E	*SE*	SG	65808	72916	72917	65855
465 160	**SE**	E	*SE*	SG	65809	72918	72919	65856
465 161	**SE**	E	*SE*	SG	65810	72920	72921	65857
465 162	**SE**	E	*SE*	SG	65811	72922	72923	65858
465 163	**SE**	E	*SE*	SG	65812	72924	72925	65859
465 164	**SE**	E	*SE*	SG	65813	72926	72927	65860
465 165	**SE**	E	*SE*	SG	65814	72928	72929	65861
465 166	**SE**	E	*SE*	SG	65815	72930	72931	65862
465 167	**SE**	E	*SE*	SG	65816	72932	72933	65863
465 168	**SE**	E	*SE*	SG	65817	72934	72935	65864

465 169	**SE**	E	*SE*	SG	65818	72936	72937	65865
465 170	**SE**	E	*SE*	SG	65819	72938	72939	65866
465 171	**SE**	E	*SE*	SG	65820	72940	72941	65867
465 172	**SE**	E	*SE*	SG	65821	72942	72943	65868
465 173	**SE**	E	*SE*	SG	65822	72944	72945	65869
465 174	**SE**	E	*SE*	SG	65823	72946	72947	65870
465 175	**SE**	E	*SE*	SG	65824	72948	72949	65871
465 176	**SE**	E	*SE*	SG	65825	72950	72951	65872
465 177	**SE**	E	*SE*	SG	65826	72952	72953	65873
465 178	**SE**	E	*SE*	SG	65827	72954	72955	65874
465 179	**SE**	E	*SE*	SG	65828	72956	72957	65875
465 180	**SE**	E	*SE*	SG	65829	72958	72959	65876
465 181	**SE**	E	*SE*	SG	65830	72960	72961	65877
465 182	**SE**	E	*SE*	SG	65831	72962	72963	65878
465 183	**SE**	E	*SE*	SG	65832	72964	72965	65879
465 184	**SE**	E	*SE*	SG	65833	72966	72967	65880
465 185	**SE**	E	*SE*	SG	65834	72968	72969	65881
465 186	**SE**	E	*SE*	SG	65835	72970	72971	65882
465 187	**SE**	E	*SE*	SG	65836	72972	72973	65883
465 188	**SE**	E	*SE*	SG	65837	72974	72975	65884
465 189	**SE**	E	*SE*	SG	65838	72976	72977	65885
465 190	**SE**	E	*SE*	SG	65839	72978	72979	65886
465 191	**SE**	E	*SE*	SG	65840	72980	72981	65887
465 192	**SE**	E	*SE*	SG	65841	72982	72983	65888
465 193	**SE**	E	*SE*	SG	65842	72984	72985	65889
465 194	**SE**	E	*SE*	SG	65843	72986	72987	65890
465 195	**SE**	E	*SE*	SG	65844	72988	72989	65891
465 196	**SE**	E	*SE*	SG	65845	72990	72991	65892
465 197	**SE**	E	*SE*	SG	65846	72992	72993	65893

Class 465/2. Built by Metro-Cammell. **Dimensions:** 20.80/20.15 x 2.81 m.

465 235	**SE**	A	*SE*	SG	65734	72787	72788	65784
465 236	**CN**	A	*SE*	SG	65735	72789	72790	65785
465 237	**SE**	A	*SE*	SG	65736	72791	72792	65786
465 238	**SE**	A	*SE*	SG	65737	72793	72794	65787
465 239	**CN**	A	*SE*	SG	65738	72795	72796	65788
465 240	**SE**	A	*SE*	SG	65739	72797	72798	65789
465 241	**SE**	A	*SE*	SG	65740	72799	72800	65790
465 242	**SE**	A	*SE*	SG	65741	72801	72802	65791
465 243	**SE**	A	*SE*	SG	65742	72803	72804	65792
465 244	**SE**	A	*SE*	SG	65743	72805	72806	65793
465 245	**SE**	A	*SE*	SG	65744	72807	72808	65794
465 246	**CN**	A	*SE*	SG	65745	72809	72810	65795
465 247	**SE**	A	*SE*	SG	65746	72811	72812	65796
465 248	**SE**	A	*SE*	SG	65747	72813	72814	65797
465 249	**CN**	A	*SE*	SG	65748	72815	72816	65798
465 250	**SE**	A	*SE*	SG	65749	72817	72818	65799

Class 465/9. Built by Metro-Cammell. Refurbished 2005 for longer distance services, with the addition of First Class. Details as Class 465/0 unless stated.
Formation: DMCO–TSO(A)–TSO(B)–DMCO.
Seating Layout: 1: 2+2 facing/unidirectional, 2: 3+2 facing/unidirectional.

65700–733. DMCO(A). Lot No. 31103 Metro-Cammell 1991–93. 12/68. 39.2t.
72719–785 (odd nos.) TSO(A). Lot No. 31104 Metro-Cammell 1991–92. –/76 1T 2W. 30.3t.
72720–786 (even nos.) TSO(B). Lot No. 31105 Metro-Cammell 1991–92. –/90. 29.5t.
65750–783. DMCO(B). Lot No. 31103 Metro-Cammell 1991–93. 12/68. 39.2t.

465 901	(465 201)	**CN**	A	*SE*	SG	65700	72719 72720	65750
465 902	(465 202)	**SE**	A	*SE*	SG	65701	72721 72722	65751
465 903	(465 203)	**SE**	A	*SE*	SG	65702	72723 72724	65752
465 904	(465 204)	**SE**	A	*SE*	SG	65703	72725 72726	65753
465 905	(465 205)	**SE**	A	*SE*	SG	65704	72727 72728	65754
465 906	(465 206)	**CN**	A	*SE*	SG	65705	72729 72730	65755
465 907	(465 207)	**SE**	A	*SE*	SG	65706	72731 72732	65756
465 908	(465 208)	**CN**	A	*SE*	SG	65707	72733 72734	65757
465 909	(465 209)	**CN**	A	*SE*	SG	65708	72735 72736	65758
465 910	(465 210)	**SE**	A	*SE*	SG	65709	72737 72738	65759
465 911	(465 211)	**CN**	A	*SE*	SG	65710	72739 72740	65760
465 912	(465 212)	**SE**	A	*SE*	SG	65711	72741 72742	65761
465 913	(465 213)	**SE**	A	*SE*	SG	65712	72743 72744	65762
465 914	(465 214)	**SE**	A	*SE*	SG	65713	72745 72746	65763
465 915	(465 215)	**SE**	A	*SE*	SG	65714	72747 72748	65764
465 916	(465 216)	**SE**	A	*SE*	SG	65715	72749 72750	65765
465 917	(465 217)	**SE**	A	*SE*	SG	65716	72751 72752	65766
465 918	(465 218)	**SE**	A	*SE*	SG	65717	72753 72754	65767
465 919	(465 219)	**CN**	A	*SE*	SG	65718	72755 72756	65768
465 920	(465 220)	**SE**	A	*SE*	SG	65719	72757 72758	65769
465 921	(465 221)	**SE**	A	*SE*	SG	65720	72759 72760	65770
465 922	(465 222)	**SE**	A	*SE*	SG	65721	72761 72762	65771
465 923	(465 223)	**SE**	A	*SE*	SG	65722	72763 72764	65772
465 924	(465 224)	**SE**	A	*SE*	SG	65723	72765 72766	65773
465 925	(465 225)	**SE**	A	*SE*	SG	65724	72767 72768	65774
465 926	(465 226)	**CN**	A	*SE*	SG	65725	72769 72770	65775
465 927	(465 227)	**SE**	A	*SE*	SG	65726	72771 72772	65776
465 928	(465 228)	**SE**	A	*SE*	SG	65727	72773 72774	65777
465 929	(465 229)	**SE**	A	*SE*	SG	65728	72775 72776	65778
465 930	(465 230)	**SE**	A	*SE*	SG	65729	72777 72778	65779
465 931	(465 231)	**SE**	A	*SE*	SG	65730	72779 72780	65780
465 932	(465 232)	**SE**	A	*SE*	SG	65731	72781 72782	65781
465 933	(465 233)	**SE**	A	*SE*	SG	65732	72783 72784	65782
465 934	(465 234)	**CN**	A	*SE*	SG	65733	72785 72786	65783

Name: 465903 Remembrance

CLASS 466   NETWORKER   GEC-ALSTHOM

Inner/outer suburban units.

Formation: DMSO–DTSO.
Construction: Welded aluminium alloy.
Traction Motors: Two GEC-Alsthom G352AY asynchronous of 280 kW.
Wheel Arrangement: Bo-Bo + 2-2.   **Couplers:** Tightlock.
Braking: Disc, rheostatic & regen.   **Control System:** 1992-type GTO Inverter.
Dimensions: 20.80 x 2.80 m.   **Maximum Speed:** 75 mph.
Bogies: BREL P3/T3.   **Doors:** Sliding plug.
Gangways: Within unit.   **Seating Layout:** 3+2 facing/unidirectional.
Multiple Working: Within class and with Class 465.

DMSO. Lot No. 31128 Birmingham 1993–94. –/86. 40.6 t.
DTSO. Lot No. 31129 Birmingham 1993–94. –/82 1T. 31.4 t.

466 001	**SE**	A	*SE*	SG	64860	78312
466 002	**SE**	A	*SE*	SG	64861	78313
466 003	**SE**	A	*SE*	SG	64862	78314
466 004	**SE**	A	*SE*	SG	64863	78315
466 005	**SE**	A	*SE*	SG	64864	78316
466 006	**SE**	A	*SE*	SG	64865	78317
466 007	**SE**	A	*SE*	SG	64866	78318
466 008	**SE**	A	*SE*	SG	64867	78319
466 009	**SE**	A	*SE*	SG	64868	78320
466 010	**SE**	A	*SE*	SG	64869	78321
466 011	**SE**	A	*SE*	SG	64870	78322
466 012	**SE**	A	*SE*	SG	64871	78323
466 013	**SE**	A	*SE*	SG	64872	78324
466 014	**SE**	A	*SE*	SG	64873	78325
466 015	**SE**	A	*SE*	SG	64874	78326
466 016	**SE**	A	*SE*	SG	64875	78327
466 017	**SE**	A	*SE*	SG	64876	78328
466 018	**SE**	A	*SE*	SG	64877	78329
466 019	**SE**	A	*SE*	SG	64878	78330
466 020	**SE**	A	*SE*	SG	64879	78331
466 021	**SE**	A	*SE*	SG	64880	78332
466 022	**SE**	A	*SE*	SG	64881	78333
466 023	**SE**	A	*SE*	SG	64882	78334
466 024	**SE**	A	*SE*	SG	64883	78335
466 025	**SE**	A	*SE*	SG	64884	78336
466 026	**SE**	A	*SE*	SG	64885	78337
466 027	**SE**	A	*SE*	SG	64886	78338
466 028	**SE**	A	*SE*	SG	64887	78339
466 029	**SE**	A	*SE*	SG	64888	78340
466 030	**SE**	A	*SE*	SG	64889	78341
466 031	**SE**	A	*SE*	SG	64890	78342
466 032	**SE**	A	*SE*	SG	64891	78343
466 033	**SE**	A	*SE*	SG	64892	78344
466 034	**SE**	A	*SE*	SG	64893	78345
466 035	**SE**	A	*SE*	SG	64894	78346

466036	**SE**	A	*SE*	SG	64895	78347
466037	**SE**	A	*SE*	SG	64896	78348
466038	**SE**	A	*SE*	SG	64897	78349
466039	**SE**	A	*SE*	SG	64898	78350
466040	**SE**	A	*SE*	SG	64899	78351
466041	**SE**	A	*SE*	SG	64900	78352
466042	**SE**	A	*SE*	SG	64901	78353
466043	**SE**	A	*SE*	SG	64902	78354

CLASS 483   METRO-CAMMELL

Built 1938 onwards for LTE. Converted 1989–90 for the Isle of Wight Line.

Formation: DMSO–DMSO.
System: 660 V DC third rail.
Construction: Steel.
Traction Motors: Two Crompton Parkinson/GEC/BTH LT100 of 125 kW.
Braking: Tread.   **Dimensions:** 16.15 x 2.69 m.
Bogies: LT design.   **Couplers:** Wedglock.
Gangways: None. End doors.
Control System: Pneumatic Camshaft Motor (PCM).
Doors: Sliding.   **Maximum Speed:** 45 mph.
Seating Layout: Longitudinal or 2+2 facing/unidirectional.
Multiple Working: Within class.
Notes: The last three numbers of the unit number only are carried.

Former London Underground numbers are shown in parentheses.

DMSO (A). Lot No. 31071. –/40. 27.4 t.
DMSO (B). Lot No. 31072. –/42. 27.4 t.

483002	**LT**	SW		RY	122	(10221)	225	(11142)	RAPTOR
483004	**LT**	SW	*SW*	RY	124	(10205)	224	(11205)	
483006	**LT**	SW	*SW*	RY	126	(10297)	226	(11297)	
483007	**LT**	SW	*SW*	RY	127	(10291)	227	(11291)	
483008	**LT**	SW	*SW*	RY	128	(10255)	228	(11255)	
483009	**LT**	SW	*SW*	RY	129	(10229)	229	(11229)	

CLASS 507   BREL YORK

Formation: BDMSO–TSO–DMSO.
Construction: Steel underframe, aluminium alloy body and roof.
Traction Motors: Four GEC G310AZ of 82.125 kW.
Wheel Arrangement: Bo-Bo + 2-2 + Bo-Bo.
Braking: Disc & rheostatic.   **Dimensions:** 20.18 x 2.82 m.
Bogies: BX1.   **Couplers:** Tightlock.
Gangways: Within unit + end doors. **Control System:** Camshaft.
Doors: Sliding.   **Maximum Speed:** 75 mph.
Seating Layout: All refurbished with 2+2 high-back facing seating.
Multiple Working: Within class and with Class 508.

Advertising livery: 507 002 Liverpool Hope University (white).

BDMSO. Lot No. 30906 1978–80. –/56(3) 1W. 37.0 t.
TSO. Lot No. 30907 1978–80. –/74. 25.5 t.
DMSO. Lot No. 30908 1978–80. –/56(3) 1W. 35.5 t.

507 001	**ME**	A	*ME*	BD	64367	71342	64405
507 002	**AL**	A	*ME*	BD	64368	71343	64406
507 003	**MY**	A	*ME*	BD	64369	71344	64407
507 004	**ME**	A	*ME*	BD	64388	71345	64408
507 005	**MY**	A	*ME*	BD	64371	71346	64409
507 006	**MY**	A	*ME*	BD	64372	71347	64410
507 007	**ME**	A	*ME*	BD	64373	71348	64411
507 008	**ME**	A	*ME*	BD	64374	71349	64412
507 009	**ME**	A	*ME*	BD	64375	71350	64413
507 010	**MY**	A	*ME*	BD	64376	71351	64414
507 011	**MY**	A	*ME*	BD	64377	71352	64415
507 012	**ME**	A	*ME*	BD	64378	71353	64416
507 013	**ME**	A	*ME*	BD	64379	71354	64417
507 014	**MY**	A	*ME*	BD	64380	71355	64418
507 015	**MY**	A	*ME*	BD	64381	71356	64419
507 016	**MY**	A	*ME*	BD	64382	71357	64420
507 017	**MY**	A	*ME*	BD	64383	71358	64421
507 018	**MY**	A	*ME*	BD	64384	71359	64422
507 019	**ME**	A	*ME*	BD	64385	71360	64423
507 020	**MY**	A	*ME*	BD	64386	71361	64424
507 021	**MY**	A	*ME*	BD	64387	71362	64425
507 023	**ME**	A	*ME*	BD	64389	71364	64427
507 024	**ME**	A	*ME*	BD	64390	71365	64428
507 025	**MY**	A	*ME*	BD	64391	71366	64429
507 026	**MY**	A	*ME*	BD	64392	71367	64430
507 027	**MY**	A	*ME*	BD	64393	71368	64431
507 028	**ME**	A	*ME*	BD	64394	71369	64432
507 029	**ME**	A	*ME*	BD	64395	71370	64433
507 030	**MY**	A	*ME*	BD	64396	71371	64434
507 031	**MY**	A	*ME*	BD	64397	71372	64435
507 032	**ME**	A	*ME*	BD	64398	71373	64436
507 033	**MY**	A	*ME*	BD	64399	71374	64437

Names:

507 004 Bob Paisley
507 008 Harold Wilson
507 009 Dixie Dean
507 023 Operations Inspector Stuart Mason

▲ London Overground 378 139 arrives at Haggerston with the 15.32 Highbury & Islington–Clapham Junction East London Line service on 26/07/13. **Robert Pritchard**

▼ Still in National Express livery, Greater Anglia 379 007 and 379 026 approach Cheshunt with the 11.45 Stansted Airport–London Liverpool Street on 21/07/14.
Antony Guppy

▲ ScotRail 380 101 calls at Carluke with the 13.48 Glasgow Central–Edinburgh Waverley on 12/03/14.   **Robin Ralston**

▼ Virgin Trains 11-car Pendolino 390 141 arrives at Rugby with the 12.33 Birmingham New Street–London Euston on 17/07/13.   **Stewart Armstrong**

▲ Southeastern High Speed 395 022 leaves Ebbsfleet International with the 14.53 Margate–London St Pancras on 08/09/14. **Antony Guppy**

▼ Gatwick Express-liveried 442 417 and 442 419 pass Coulsdon South with the 16.00 London Victoria–Gatwick Airport on 15/03/12. **Robert Pritchard**

▲ SWT white-liveried 444 016 passes Eastleigh with the 13.35 London Waterloo–Weymouth on 28/07/14. **Stewart Armstrong**

▼ SWT blue-liveried 450 076 passes Battledown working the 11.39 London Waterloo–Poole on 14/01/14. **Chris Wilson**

▲ Southern-liveried 455 822 leaves Mitcham Junction with the 10.47 London Victoria–Epsom on 12/07/14. **Chris Wilson**

▼ Refurbished and repainted into the South West Trains red suburban livery, 456 003 stands at Wimbledon depot on 30/06/14. **Brian Garvin**

▲ The rebuilt 5-car 458/5s slowly started to enter traffic during 2014. On 07/03/14 458 531 and 458 534 pass Vauxhall with an empty stock working from London Waterloo to Wimbledon depot following a press launch. **Robert Pritchard**

▼ Southeastern-liveried 465 042 leads a Charing Cross–Dartford service into Lewisham on 25/01/14. **Andrew Mason**

▲ A small fleet of former London Underground 1938 Stock units are used on the Isle of Wight railway, operated by South West Trains. On 18/05/14 483 006 leaves Smallbrook Junction with the 14.38 Shanklin–Ryde Pier Head.   **Nigel Gibbs**

▼ New Merseyrail-liveried 507 031 leads 507 002 into Brunswick with the 12.28 Southport–Hunts Cross as 508 127 heads north on 07/09/14.   **Robert Pritchard**

▲ The first of the new Thameslink Class 700s was due to arrive in Britain in 2015 following testing. On 04/09/14 12-car unit 700 101 stands at Velim station in the Czech Republic on its way to the nearby test circuit. **Quintus Vosman**

▼ One of the ten new Siemens Class 374 Eurostar Velaros, 4005/06, is seen at Mitry-Mory in France on its way to the LGV Rhin-Rhône line for tests on 24/06/14. These trains will enter traffic in 2015. **Christophe Masse**

CLASS 508   BREL YORK

Formation: DMSO–TSO–BDMSO.
Construction: Steel underframe, aluminium alloy body and roof.
Traction Motors: Four GEC G310AZ of 82.125 kW.
Wheel Arrangement: Bo-Bo + 2-2 + Bo-Bo.
Braking: Disc & rheostatic.   **Dimensions:** 20.18 x 2.82 m.
Bogies: BX1.   **Couplers:** Tightlock.
Gangways: Within unit + end doors. **Control System:** Camshaft.
Doors: Sliding.   **Maximum Speed:** 75 mph.
Seating Layout: All refurbished with 2+2 high-back facing seating.
Multiple Working: Within class and with Class 507.

Advertising livery: 508111 Beatles Story (blue).

DMSO. Lot No. 30979 1979–80. –/56(3) 1W. 36.0 t.
TSO. Lot No. 30980 1979–80. –/74. 26.5 t.
BDMSO. Lot No. 30981 1979–80. –/56(3) 1W. 36.5 t.

508 103	**ME**	A	*ME*	BD	64651	71485	64694
508 104	**ME**	A	*ME*	BD	64652	71486	64695
508 108	**MY**	A	*ME*	BD	64656	71490	64699
508 110	**ME**	A	*ME*	BD	64658	71492	64701
508 111	**AL**	A	*ME*	BD	64659	71493	64702
508 112	**ME**	A	*ME*	BD	64660	71494	64703
508 114	**MY**	A	*ME*	BD	64662	71496	64705
508 115	**MY**	A	*ME*	BD	64663	71497	64706
508 117	**MY**	A	*ME*	BD	64665	71499	64708
508 120	**MY**	A	*ME*	BD	64668	71502	64711
508 122	**MY**	A	*ME*	BD	64670	71504	64713
508 123	**MY**	A	*ME*	BD	64671	71505	64714
508 124	**MY**	A	*ME*	BD	64672	71506	64715
508 125	**ME**	A	*ME*	BD	64673	71507	64716
508 126	**MY**	A	*ME*	BD	64674	71508	64717
508 127	**ME**	A	*ME*	BD	64675	71509	64718
508 128	**MY**	A	*ME*	BD	64676	71510	64719
508 130	**MY**	A	*ME*	BD	64678	71512	64721
508 131	**ME**	A	*ME*	BD	64679	71513	64722
508 134	**ME**	A	*ME*	BD	64682	71516	64725
508 136	**MY**	A	*ME*	BD	64684	71518	64727
508 137	**ME**	A	*ME*	BD	64685	71519	64728
508 138	**ME**	A	*ME*	BD	64686	71520	64729
508 139	**MY**	A	*ME*	BD	64687	71521	64730
508 140	**ME**	A	*ME*	BD	64688	71522	64731
508 141	**ME**	A	*ME*	BD	64689	71523	64732
508 143	**ME**	A	*ME*	BD	64691	71525	64734

Names:

508 123 William Roscoe
508 136 Wilfred Owen MC

3. DUAL VOLTAGE THAMESLINK UNITS

The large fleet of EMUs currently under construction for the Thameslink routes have been designated Class 700. The first of these units should be delivered for testing in summer 2014 and the fleet will enter traffic between early 2016 and late 2018. They are being financed by Cross London Trains (a consortium of Siemens Project Ventures, Innisfree Ltd and 3i Infrastructure Ltd) and will be based at new depots being constructed at Three Bridges and Hornsey. The units will have 6-digit vehicle numbers. Full details awaited.

CLASS 700   DESIRO CITY   SIEMENS

Formations (8-car): DMCO–PTSO–MSO–TSO–TSO–MSO–PTSO–DMCO
or **(12-car):** DMCO–PTSO–MSO–MSO–TSO–TSO–TSO–TSO–MSO–MSO–PTSO–DMCO.
Systems: 25 kV AC overhead/750 V DC third rail.
Construction: Aluminium.
Traction Motors:
Wheel Arrangement (8-car): Bo-Bo + 2-2 + Bo-Bo + 2-2 + 2-2 + Bo-Bo + 2-2 + Bo-Bo. **(12-car):** Bo-Bo + 2-2 + Bo-Bo + Bo-Bo + 2-2 + 2-2 + 2-2 + 2-2 + Bo-Bo + Bo-Bo + 2-2 + Bo-Bo.
Braking: **Dimensions:** 20.0 m x 2.80 m.
Bogies: Siemens SF7000 inside-frame. **Couplers:**
Gangways: Within unit.   **Control System:** IGBT Inverter.
Doors: **Maximum Speed:** 100 mph.
Heating & ventilation: Air conditioning.
Seating Layout: 2+2 facing/unidirectional.
Multiple Working: Within class.

Class 700/0. 8-car units.

DMCO(A). Siemens Krefeld 2014–18. 26/16(3).   t.
PTSO. Siemens Krefeld 2014–18. –/54 1T.   t.
MSO. Siemens Krefeld 2014–18. –/64.   t.
TSO. Siemens Krefeld 2014–18. –/56.   t.
TSO. Siemens Krefeld 2014–18. –/40(8) 1TD 2W.   t.
MSO. Siemens Krefeld 2014–18. –/64.   t.
PTSO. Siemens Krefeld 2014–18. –/54 1T.   t.
DMCO(B). Siemens Krefeld 2014–18. 26/16(3).   t.

700 001	401001	402001	403001	406001
	407001	410001	411001	412001
700 002	401002	402002	403002	406002
	407002	410002	411002	412002
700 003	401003	402003	403003	406003
	407003	410003	411003	412003
700 004	401004	402004	403004	406004
	407004	410004	411004	412004

700 005	401005	402005	403005	406005
	407005	410005	411005	412005
700 006	401006	402006	403006	406006
	407006	410006	411006	412006
700 007	401007	402007	403007	406007
	407007	410007	411007	412007
700 008	401008	402008	403008	406008
	407008	410008	411008	412008
700 009	401009	402009	403009	406009
	407009	410009	411009	412009
700 010	401010	402010	403010	406010
	407010	410010	411010	412010
700 011	401011	402011	403011	406011
	407011	410011	411011	412011
700 012	401012	402012	403012	406012
	407012	410012	411012	412012
700 013	401013	402013	403013	406013
	407013	410013	411013	412013
700 014	401014	402014	403014	406014
	407014	410014	411014	412014
700 015	401015	402015	403015	406015
	407015	410015	411015	412015
700 016	401016	402016	403016	406016
	407016	410016	411016	412016
700 017	401017	402017	403017	406017
	407017	410017	411017	412017
700 018	401018	402018	403018	406018
	407018	410018	411018	412018
700 019	401019	402019	403019	406019
	407019	410019	411019	412019
700 020	401020	402020	403020	406020
	407020	410020	411020	412020
700 021	401021	402021	403021	406021
	407021	410021	411021	412021
700 022	401022	402022	403022	406022
	407022	410022	411022	412022
700 023	401023	402023	403023	406023
	407023	410023	411023	412023
700 024	401024	402024	403024	406024
	407024	410024	411024	412024
700 025	401025	402025	403025	406025
	407025	410025	411025	412025
700 026	401026	402026	403026	406026
	407026	410026	411026	412026
700 027	401027	402027	403027	406027
	407027	410027	411027	412027
700 028	401028	402028	403028	406028
	407028	410028	411028	412028
700 029	401029	402029	403029	406029
	407029	410029	411029	412029

700 030	401030	402030	403030	406030
	407030	410030	411030	412030
700 031	401031	402031	403031	406031
	407031	410031	411031	412031
700 032	401032	402032	403032	406032
	407032	410032	411032	412032
700 033	401033	402033	403033	406033
	407033	410033	411033	412033
700 034	401034	402034	403034	406034
	407034	410034	411034	412034
700 035	401035	402035	403035	406035
	407035	410035	411035	412035
700 036	401036	402036	403036	406036
	407036	410036	411036	412036
700 037	401037	402037	403037	406037
	407037	410037	411037	412037
700 038	401038	402038	403038	406038
	407038	410038	411038	412038
700 039	401039	402039	403039	406039
	407039	410039	411039	412039
700 040	401040	402040	403040	406040
	407040	410040	411040	412040
700 041	401041	402041	403041	406041
	407041	410041	411041	412041
700 042	401042	402042	403042	406042
	407042	410042	411042	412042
700 043	401043	402043	403043	406043
	407043	410043	411043	412043
700 044	401044	402044	403044	406044
	407044	410044	411044	412044
700 045	401045	402045	403045	406045
	407045	410045	411045	412045
700 046	401046	402046	403046	406046
	407046	410046	411046	412046
700 047	401047	402047	403047	406047
	407047	410047	411047	412047
700 048	401048	402048	403048	406048
	407048	410048	411048	412048
700 049	401049	402049	403049	406049
	407049	410049	411049	412049
700 050	401050	402050	403050	406050
	407050	410050	411050	412050
700 051	401051	402051	403051	406051
	407051	410051	411051	412051
700 052	401052	402052	403052	406052
	407052	410052	411052	412052
700 053	401053	402053	403053	406053
	407053	410053	411053	412053
700 054	401054	402054	403054	406054
	407054	410054	411054	412054

700 055	401055	402055	403055	406055
	407055	410055	411055	412055
700 056	401056	402056	403056	406056
	407056	410056	411056	412056
700 057	401057	402057	403057	406057
	407057	410057	411057	412057
700 058	401058	402058	403058	406058
	407058	410058	411058	412058
700 059	401059	402059	403059	406059
	407059	410059	411059	412059
700 060	401060	402060	403060	406060
	407060	410060	411060	412060

Class 700/1. 12-car units.

DMCO(A). Siemens Krefeld 2013–18. 26/20.   t.
PTSO. Siemens Krefeld 2013–18. –/54 1T.   t.
MSO. Siemens Krefeld 2013–18. –/60 (3).   t.
MSO. Siemens Krefeld 2013–18. –/56 1T.   t.
TSO. Siemens Krefeld 2013–18. –/64.   t.
TSO. Siemens Krefeld 2013–18. –/56.   t.
TSO. Siemens Krefeld 2013–18. –/37(8) 1TD 2W.   t.
TSO. Siemens Krefeld 2013–18. –/64.   t.
MSO. Siemens Krefeld 2013–18. –/56 1T.   t.
MSO. Siemens Krefeld 2013–18. –/60(3).   t.
PTSO. Siemens Krefeld 2013–18. –/54 1T.   t.
DMCO(B). Siemens Krefeld 2013–18. 26/20.   t.

700 101	401101	402101	403101	404101	405101	406101
	407101	408101	409101	410101	411101	412101
700 102	401102	402102	403102	404102	405102	406102
	407102	408102	409102	410102	411102	412102
700 103	401103	402103	403103	404103	405103	406103
	407103	408103	409103	410103	411103	412103
700 104	401104	402104	403104	404104	405104	406104
	407104	408104	409104	410104	411104	412104
700 105	401105	402105	403105	404105	405105	406105
	407105	408105	409105	410105	411105	412105
700 106	401106	402106	403106	404106	405106	406106
	407106	408106	409106	410106	411106	412106
700 107	401107	402107	403107	404107	405107	406107
	407107	408107	409107	410107	411107	412107
700 108	401108	402108	403108	404108	405108	406108
	407108	408108	409108	410108	411108	412108
700 109	401109	402109	403109	404109	405109	406109
	407109	408109	409109	410109	411109	412109
700 110	401110	402110	403110	404110	405110	406110
	407110	408110	409110	410110	411110	412110
700 111	401111	402111	403111	404111	405111	406111
	407111	408111	409111	410111	411111	412111
700 112	401112	402112	403112	404112	405112	406112
	407112	408112	409112	410112	411112	412112

700 113	401113 402113 403113 404113 405113 406113
	407113 408113 409113 410113 411113 412113
700 114	401114 402114 403114 404114 405114 406114
	407114 408114 409114 410114 411114 412114
700 115	401115 402115 403115 404115 405115 406115
	407115 408115 409115 410115 411115 412115
700 116	401116 402116 403116 404116 405116 406116
	407116 408116 409116 410116 411116 412116
700 117	401117 402117 403117 404117 405117 406117
	407117 408117 409117 410117 411117 412117
700 118	401118 402118 403118 404118 405118 406118
	407118 408118 409118 410118 411118 412118
700 119	401119 402119 403119 404119 405119 406119
	407119 408119 409119 410119 411119 412119
700 120	401120 402120 403120 404120 405120 406120
	407120 408120 409120 410120 411120 412120
700 121	401121 402121 403121 404121 405121 406121
	407121 408121 409121 410121 411121 412121
700 122	401122 402122 403122 404122 405122 406122
	407122 408122 409122 410122 411122 412122
700 123	401123 402123 403123 404123 405123 406123
	407123 408123 409123 410123 411123 412123
700 124	401124 402124 403124 404124 405124 406124
	407124 408124 409124 410124 411124 412124
700 125	401125 402125 403125 404125 405125 406125
	407125 408125 409125 410125 411125 412125
700 126	401126 402126 403126 404126 405126 406126
	407126 408126 409126 410126 411126 412126
700 127	401127 402127 403127 404127 405127 406127
	407127 408127 409127 410127 411127 412127
700 128	401128 402128 403128 404128 405128 406128
	407128 408128 409128 410128 411128 412128
700 129	401129 402129 403129 404129 405129 406129
	407129 408129 409129 410129 411129 412129
700 130	401130 402130 403130 404130 405130 406130
	407130 408130 409130 410130 411130 412130
700 131	401131 402131 403131 404131 405131 406131
	407131 408131 409131 410131 411131 412131
700 132	401132 402132 403132 404132 405132 406132
	407132 408132 409132 410132 411132 412132
700 133	401133 402133 403133 404133 405133 406133
	407133 408133 409133 410133 411133 412133
700 134	401134 402134 403134 404134 405134 406134
	407134 408134 409134 410134 411134 412134
700 135	401135 402135 403135 404135 405135 406135
	407135 408135 409135 410135 411135 412135
700 136	401136 402136 403136 404136 405136 406136
	407136 408136 409136 410136 411136 412136
700 137	401137 402137 403137 404137 405137 406137
	407137 408137 409137 410137 411137 412137

700 138	401138 402138 403138 404138 405138 406138
	407138 408138 409138 410138 411138 412138
700 139	401139 402139 403139 404139 405139 406139
	407139 408139 409139 410139 411139 412139
700 140	401140 402140 403140 404140 405140 406140
	407140 408140 409140 410140 411140 412140
700 141	401141 402141 403141 404141 405141 406141
	407141 408141 409141 410141 411141 412141
700 142	401142 402142 403142 404142 405142 406142
	407142 408142 409142 410142 411142 412142
700 143	401143 402143 403143 404143 405143 406143
	407143 408143 409143 410143 411143 412143
700 144	401144 402144 403144 404144 405144 406144
	407144 408144 409144 410144 411144 412144
700 145	401145 402145 403145 404145 405145 406145
	407145 408145 409145 410145 411145 412145
700 146	401146 402146 403146 404146 405146 406146
	407146 408146 409146 410146 411146 412146
700 147	401147 402147 403147 404147 405147 406147
	407147 408147 409147 410147 411147 412147
700 148	401148 402148 403148 404148 405148 406148
	407148 408148 409148 410148 411148 412148
700 149	401149 402149 403149 404149 405149 406149
	407149 408149 409149 410149 411149 412149
700 150	401150 402150 403150 404150 405150 406150
	407150 408150 409150 410150 411150 412150
700 151	401151 402151 403151 404151 405151 406151
	407151 408151 409151 410151 411151 412151
700 152	401152 402152 403152 404152 405152 406152
	407152 408152 409152 410152 411152 412152
700 153	401153 402153 403153 404153 405153 406153
	407153 408153 409153 410153 411153 412153
700 154	401154 402154 403154 404154 405154 406154
	407154 408154 409154 410154 411154 412154
700 155	401155 402155 403155 404155 405155 406155
	407155 408155 409155 410155 411155 412155

4. EUROSTAR UNITS

The original Eurostar Class 373 units were built for and are normally used on services between Britain and continental Europe via the Channel Tunnel. SNCF-owned units 3203/04, 3225/26 and 3227/28 have been removed from the Eurostar pool and were used on Paris–Lille services, but have now been stored. As they are not now permitted through the Channel Tunnel they are not listed here.

Each Class 373 train consists of two 10-car units coupled, with a motor car at each driving end. All units are articulated with an extra motor bogie on the coach adjacent to the motor car.

All Class 373 sets can be used between London St Pancras and Paris, Brussels and Disneyland Paris. Certain sets (shown *) are equipped for 1500 V DC operation and are used for the winter service to Bourg Saint Maurice and the summer service to Avignon.

Seven 8-car Class 373 sets were built for Regional Eurostar services, but all except one power car (3308) and one half set are on long-term hire to SNCF for use on French internal services so are not listed here. The spare half set (from 3308/07) is stored at Temple Mills depot.

The second generation Eurostar trains, Class 374s, are currently under construction, although these will supplement rather than replace the 373s.

Formation: DM–MSO–4TSO–RB–2TFO–TBFO. Gangwayed within pair of units. Air conditioned.
Construction: Steel.
Supply Systems: 25 kV AC 50 Hz overhead or 3000 V DC overhead (* also equipped for 1500 V DC overhead operation).
Control System: GTO–GTO Inverter on UK 750V DC and 25kVAC, GTO Chopper on SNCB 3000VDC.
Wheel Arrangement: Bo-Bo + Bo–2–2–2–2–2–2–2–2–2–2.
Length: 22.15 m (DM), 21.85 m (MS & TBF), 18.70 m (other cars).
Couplers: Schaku 10S at outer ends, Schaku 10L at inner end of each DM and outer ends of each sub set.
Maximum Speed: 186 mph (300 km/h)
Built: 1992–93 by GEC-Alsthom/Brush/ANF/De Dietrich/BN Construction/ACEC.
DM vehicles carry the set numbers indicated below.

CLASS 373   ORIGINAL-BUILD UNITS

10-car sets. Built for services starting from or terminating in London Waterloo (now St Pancras). Individual vehicles in each set are allocated numbers 373xxx0 + 373xxx1 + 373xxx2 + 373xxx3 + 373xxx4 + 373xxx5 + 373xxx6 + 373xxx7 + 373xxx8 + 373xxx9, where 3xxx denotes the set number.

At the time of writing power cars 3015 and 3016 are carrying the numbers 3212 and 3211 respectively, as they are running with the trailers from set 3211/12.

373xxx0 series. DM. Lot No. 31118 1992–95. 68.5 t.
373xxx1 series. MSO. Lot No. 31119 1992–95. –/48 2T. 44.6 t.
373xxx2 series. TSO. Lot No. 31120 1992–95. –/56 1T. 28.1 t.
373xxx3 series. TSO. Lot No. 31121 1992–95. –/56 2T. 29.7 t.
373xxx4 series. TSO. Lot No. 31122 1992–95. –/56 1T. 28.3 t.
373xxx5 series. TSO. Lot No. 31123 1992–95. –/56 2T. 29.2 t.
373xxx6 series. RB. Lot No.31124 1992–95. 31.1 t.
373xxx7 series. TFO. Lot No. 31125 1992–95. 39/– 1T. 29.6 t.
373xxx8 series. TFO. Lot No. 31126 1992–95. 39/– 1T. 32.2 t.
373xxx9 series. TBFO. Lot No. 31127 1992–95. 25/– 1TD. 39.4 t.

3001	**EU**	EU	*EU*	TI		3107	**EU**	SB	*EU*	FF
3002	**EU**	EU	*EU*	TI		3108	**EU**	SB	*EU*	FF
3003	**EU**	EU	*EU*	TI		3201 *	**EU**	SF	*EU*	LY
3004	**EU**	EU	*EU*	TI		3202 *	**EU**	SF	*EU*	LY
3005	**EU**	EU	*EU*	TI		3205	**EU**	SF	*EU*	LY
3006	**EU**	EU	*EU*	TI		3206	**EU**	SF	*EU*	LY
3007	**EU**	EU	*EU*	TI		3207 *	**EU**	SF	*EU*	LY
3008	**EU**	EU	*EU*	TI		3208 *	**EU**	SF	*EU*	LY
3009	**EU**	EU	*EU*	TI		3209 *	**EU**	SF	*EU*	LY
3010	**EU**	EU	*EU*	TI		3210 *	**EU**	SF	*EU*	LY
3011	**EU**	EU	*EU*	TI		3211	**EU**	SF	*EU*	LY
3012	**EU**	EU	*EU*	TI		3212	**EU**	SF	*EU*	LY
3013	**EU**	EU	*EU*	TI		3213 *	**EU**	SF	*EU*	LY
3014	**EU**	EU	*EU*	TI		3214 *	**EU**	SF	*EU*	LY
3015	**ER**	EU	*EU*	TI		3215 *	**EU**	SF	*EU*	LY
3016	**ER**	EU	*EU*	TI		3216 *	**EU**	SF	*EU*	LY
3017	**EU**	EU	*EU*	TI		3217	**EU**	SF	*EU*	LY
3018	**EU**	EU	*EU*	TI		3218	**EU**	SF	*EU*	LY
3019	**EU**	EU	*EU*	TI		3219	**EU**	SF	*EU*	LY
3020	**EU**	EU	*EU*	TI		3220	**EU**	SF	*EU*	LY
3021	**EU**	EU	*EU*	TI		3221	**EU**	SF	*EU*	LY
3022	**EU**	EU	*EU*	TI		3222	**EU**	SF	*EU*	LY
3101	**EU**	SB		TI		3223 *	**EU**	SF	*EU*	LY
3102	**EU**	SB		TI		3224 *	**EU**	SF	*EU*	LY
3103	**EU**	SB	*EU*	FF		3229 *	**EU**	SF	*EU*	LY
3104	**EU**	SB	*EU*	FF		3230 *	**EU**	SF	*EU*	LY
3105	**EU**	SB	*EU*	FF		3231	**EU**	SF	*EU*	LY
3106	**EU**	SB	*EU*	FF		3232	**EU**	SF	*EU*	LY

Spare Regional Eurostar DM:

3308	**EU**	EU		LB

Spare DM:

3999	**EU**	EU	*EU*	TI

Names:

3001/02	Tread Lightly/Voyage Vert	3013/14	LONDON 2012
3003/04	Tri-City-Athlon 2010	3207/08	MICHEL HOLLARD
3007/08	Waterloo Sunset	3209/10	THE DA VINCI CODE
3009/10	REMEMBERING FROMELLES		

CLASS 374   SIEMENS VELARO e320

8-car sets. Under construction for Eurostar. These units are similar to the DB Class 407 ICE sets. Due to enter service in 2015 ahead of a proposed St Pancras–Amsterdam service starting in 2016. Full details awaited.

Formation (provisional): DMFO–TBFO–MFO–TSO–TSO–MSO–TSO–MSORB. Gangwayed within pair of units. Air conditioned.
Construction: Aluminium.
Supply Systems: 25 kV AC 50 Hz overhead, 15 kV AC 16.7 Hz overhead, 1500 V DC overhead and 3000 V DC overhead.
Control System:
Wheel Arrangement: Bo-Bo + 2-2 + Bo-Bo + 2-2 + 2-2 + Bo-Bo + 2-2 + Bo-Bo.
Length:
Couplers:
Maximum Speed: 200 mph (320 km/h).
Built: 2012–14 by Siemens, Krefeld, Germany.

DM vehicles carry the full 12-digit European Vehicle Numbers as indicated below. For example set 4001/02 carries the numbers 93 70 3740 011-9 + 93 70 3740 012-7 + 93 70 3740 013-5 + 93 70 3740 014-3 + 93 70 3740 015-0 + 93 70 3740 016-8 + 93 70 3740 017-6 + 93 70 3740 018-4 + 93 70 3740 028-3 + 93 70 3740 027-5 + 93 70 3740 026-7 + 93 70 3740 025-9 + 93 70 3740 024-2 + 93 70 3740 023-4 + 93 70 3740 022-6 + 93 70 3740 021-8.

93 70 3740 xx1-c series. DMFO. Siemens Krefeld 2012–14.   t.
93 70 3740 xx2-c series. TBFO. Siemens Krefeld 2012–14.   t.
93 70 3740 xx3-c series. MFO. Siemens Krefeld 2012–14.   t.
93 70 3740 xx4-c series. TSO. Siemens Krefeld 2012–14.   t.
93 70 3740 xx5-c series. TSO. Siemens Krefeld 2012–14.   t.
93 70 3740 xx6-c series. MSO. Siemens Krefeld 2012–14.   t.
93 70 3740 xx7-c series. TSO. Siemens Krefeld 2012–14.   t.
93 70 3740 xx8-c series. MSORB. Siemens Krefeld 2012–14.   t.

4001	**ER**		4011	**ER**
4002	**ER**		4012	**ER**
4003	**ER**		4013	
4004	**ER**		4014	
4005	**ER**		4015	
4006	**ER**		4016	
4007	**ER**		4017	
4008	**ER**		4018	
4009	**ER**		4019	
4010	**ER**		4020	

5. SERVICE EMUS

The following unit is used by Network Rail for ERTMS testing on the Hertford Loop. The unit has been heavily modified from its original condition, and now includes a toilet.

313121   **Y**   PC   *GB*   WN   62549   71233   62613

6. EMU VEHICLES IN INDUSTRIAL SERVICE

This list comprises EMU vehicles that have been withdrawn from active service but continue to be used in industrial service.

Cl. 390	69133	69833	Virgin Trains Training Centre, Westmere Drive, Crewe, Cheshire
Cl. 390	69633	69733	The Fire Service College, Moreton-in-Marsh, Gloucestershire
Cl. 390	69933		Safety & Accident Investigation Centre, Cranfield University, Cranfield, Bedfordshire
Cl. 508	64649	64712	Emergency Services Training Centre, Seacombe, Merseyside
Cl. 508	64681	71511 64724	The Fire Service College, Moreton-in-Marsh, Gloucestershire

7. EMUS AWAITING DISPOSAL

This list comprises vehicles awaiting disposal which are stored on the national railway network.

25 kV AC 50 Hz OVERHEAD UNITS:

Cl. 309	**RR**	WC	CS	71758		
Cl. 365	**N**	X	ZN	65919		

750 V DC THIRD RAIL UNITS:

Cl. 460	**GV**	P	ZB	67901	67907	
Cl. 460	**GV**	P	LB	67903	67908	
Cl. 508	**CN**	A	ZG	64667	64680	64723
Cl. 508	**CN**	A	ZI	64710	64720	

PLATFORM 5 MAIL ORDER

EUROPEAN HANDBOOKS

The Platform 5 European Railway Handbooks are the most comprehensive guides to the rolling stock of selected European railway administrations available. Each book lists all locomotives and railcars of the country concerned, giving details of number carried and depot allocation, together with a wealth of technical data for each class of vehicle. Each book is A5 size, thread sewn and illustrated throughout with colour photographs. The Irish book also contains details of hauled coaching stock.

EUROPEAN HANDBOOKS CURRENTLY AVAILABLE:

No. 1 Benelux Railways (2012) £20.95
No. 2A German Railways Part 1:
    DB Locomotives & Multiple Units (2013) £22.95
No. 2B German Railways Part 2:
    Private Operators, Preserved & Museums ...NEW EDITION IN PREPARATION
No. 3 Austrian Railways (2012)..£19.95
No. 4 French Railways (2011) £19.95
No. 5 Swiss Railways (2009)..£19.50
No. 6 Italian Railways....................................NEW EDITION IN PREPARATION
No. 7 Irish Railways (2013) £15.95

Please add postage: 10% UK, 20% Europe, 30% Rest of World.

Telephone, fax or send your order to the Platform 5 Mail Order Department. See inside back cover of this book for details.

8. CODES

8.1. LIVERY CODES

1   "One" (metallic grey with a broad black bodyside stripe. White National Express/Greater Anglia "interim" stripe as branding).

AL   Advertising/promotional livery (see class heading for details).

CN   Connex/Southeastern (white with black window surrounds & grey lower band).

ER   Revised Eurostar (deep blue & two-tone grey).

EU   Eurostar (white with dark blue & yellow stripes).

FB   First Group dark blue.

FT   First TransPennine Express "Dynamic Lines" (varying blue with multi-coloured lines).

FU   First Group "Urban Lights" (varying blue or uniform indigo blue with pink, white & blue markings on the lower bodyside).

GA   Abellio Greater Anglia (white with red doors & black window surrounds).

GV   Gatwick Express EMU (red, white & indigo blue with mauve & blue doors).

HC   Heathrow Connect (grey with a broad deep blue bodyside band & orange doors).

HE   Heathrow Express (silver with purple doors and black window surrounds). Red advertising for Vodaphone.

LM   London Midland (grey & green with broad black stripe around the windows).

LO   London Overground (all over white with a blue solebar & black window surrounds).

LT   London Transport maroon & cream.

ME   Merseyrail (metallic silver with yellow doors).

MY   New Merseyrail (all over yellow or all over grey (alternate sides)).

N   BR Network SouthEast (white & blue with red lower bodyside stripe, grey solebar & cab ends).

NB   Northern all over dark blue.

NC   National Express white (white with blue doors).

NO   Northern (deep blue, purple & white).

NX   National Express (white with grey ends).

O   Non-standard (see class heading for details).

RM   Royal Mail (red with yellow stripes above solebar).

RR   Regional Railways (dark blue/grey with light blue & white stripes, three narrow dark blue stripes at cab ends).

SB   Southeastern High Speed (all over blue with black window surrounds).

SC   Strathclyde PTE (carmine & cream lined out in black & gold).

SD   South West Trains outer suburban livery {Class 450 style} (deep blue with red doors & orange & red cab sides).

SE   Southeastern (all over white with black window surrounds, light blue doors and (on some units) dark blue lower bodyside stripe).

SN   Southern (white & dark green with light green semi-circles at one end of each vehicle. Light grey band at solebar level).

SR   ScotRail – Scotland's Railways (dark blue with Scottish Saltire flag & white/light blue flashes).

SS South West Trains inner suburban {Class 455} (red with blue & orange flashes at unit ends).
ST Stagecoach {long-distance stock} (white & dark blue with dark blue window surrounds and red & orange swishes at unit ends).
TG Govia Thameslink interim {Class 387} (white with dark green doors}.
TL Govia Thameslink (light grey & white with light blue doors).
VT Virgin Trains silver (silver, with black window surrounds, white cantrail stripe and red roof. Red swept down at unit ends).
Y Network Rail yellow.
YR West Yorkshire PTE/Northern EMUs (red, lilac & grey).

8.2. OWNER CODES

A   Angel Trains
E   Eversholt Rail (UK)
EU Eurostar (UK)
HE British Airports Authority
MQ Macquarie Group
P   Porterbrook Leasing Company
PC Pamplona Capital Management
QW QW Rail Leasing
RM Royal Mail
SB SNCB/NMBS (Société Nationale des Chemins de fer Belges Nationale Maatschappij der Belgische Spoorwegen)
SF SNCF (Société Nationale des Chemins de fer Français)
SW South West Trains
WC West Coast Railway Company

8.3. OPERATOR CODES

C2   c2c
DB   DB Schenker
EU   Eurostar (UK)
GA   Abellio Greater Anglia
GT   Govia Thameslink Railway
HC   Heathrow Connect
HE   Heathrow Express
LM   London Midland
LO   London Overground
ME   Merseyrail
NO   Northern
SE   Southeastern
SN   Southern
SR   ScotRail
SW   South West Trains
VW   Virgin Trains

8.4. ALLOCATION & LOCATION CODES

Code	Location	Depot Operator
AD	Ashford (Kent)	Hitachi
BD	Birkenhead North	Merseyrail
BF	Bedford Cauldwell Walk	Govia Thameslink Railway
BI	Brighton Lovers Walk	Southern
CE	Crewe International	DB Schenker Rail (UK)
CS	Carnforth	West Coast Railway Company
EM	East Ham (London)	c2c
FF	Forest (Brussels)	SNCB/NMBS
GW	Glasgow Shields Road	ScotRail
HE	Hornsey (London)	Govia Thameslink Railway
IL	Ilford (London)	Abellio Greater Anglia
LB	Loughborough Works	Wabtec Rail
LG	Longsight (Manchester)	Northern
LY	Le Landy (Paris)	SNCF
MA	Manchester Longsight	Alstom
NG	New Cross Gate (London)	London Overground
NL	Neville Hill (Leeds)	East Midlands Trains/Northern
NN	Northampton King's Heath	Siemens
NT	Northam (Southampton)	Siemens
OH	Old Oak Common Heathrow (London)	Heathrow Express
RM	Ramsgate	Southeastern
RY	Ryde (Isle of Wight)	South West Trains
SG	Slade Green (London)	Southeastern
SL	Stewarts Lane (London)	Southern/Belmond
SO	Soho (Birmingham)	London Midland
SU	Selhurst (Croydon)	Southern
TI	Temple Mills (London)	Eurostar
WB	Wembley (London)	Alstom
WD	Wimbledon (London)	South West Trains
ZA	RTC Business Park (Derby)	Railway Vehicle Engineering
ZB	Doncaster Works	Wabtec Rail
ZC	Crewe Works	Bombardier Transportation UK
ZD	Derby Works	Bombardier Transportation UK
ZG	Eastleigh Works	Arlington Fleet Services
ZH	Springburn Depot (Glasgow)	Knorr-Bremse Rail Systems (UK)
ZI	Ilford Works	Bombardier Transportation UK
ZJ	Stoke-on-Trent Works	Axiom Rail (Stoke)
ZK	Kilmarnock Works	Wabtec Rail Scotland
ZN	Wolverton Works	Knorr-Bremse Rail Systems (UK)
ZR	York (Holgate Works)	Network Rail